# ENERGY
# ALTERNATIVES

For information on and a full description of any of the
Time-Life Books series listed above, please call
1-800-621-7026 or write:
Reader Information
Time-Life Customer Service
P.O. Box C-32068
Richmond, Virginia 23261-2068

This volume is part of a series offering homeowners
detailed instructions on repairs, construction
and improvements they can undertake themselves.

HOME REPAIR
AND IMPROVEMENT

# ENERGY
# ALTERNATIVES

BY THE EDITORS OF
TIME-LIFE BOOKS

TIME-LIFE BOOKS
ALEXANDRIA, VIRGINIA

**HOME REPAIR AND IMPROVEMENT**

Editor | John Paul Porter
Senior Editor | Betsy Frankel
Designer | Edward Frank

Editorial Staff for Energy Alternatives

Text Editors | Rachel Cox, Brooke Stoddard (principals), Lynn R. Addison, Tim Appenzeller, Robert Doyle, Victoria W. Monks, Mark M. Steele, William Worsley
Writers | Kevin D. Armstrong, Carol Jane Corner, Leon Greene, Kathleen M. Kiely, Brian McGinn, Kirk Y. Saunders
Art Associates | George Bell, Fred Holz, Lorraine D. Rivard, Peter C. Simmons
Picture Coordinator | Betsy Donahue
Editorial Assistant | Cathy Sharpe
Special Contributors | William Doyle, Lydia Preston

Editorial Operations

Copy Chief | Diane Ullius
Editorial Operations Manager | Caroline A. Boubin
Production | Celia Beattie
Quality Control | James J. Cox (director)
Library | Louise D. Forstall

Correspondents: Elisabeth Kraemer-Singh (Bonn); Maria Vincenza Aloisi (Paris); Ann Natanson (Rome).

THE CONSULTANTS: Roswell Ard is a consulting structural engineer and a professional home inspector in northern Michigan. Trained as a mechanical and electrical engineer, he has worked as a design consultant specializing in automatic controls for industrial machinery.

Energy Associates is a home-improvement contracting company specializing in energy construction and solar applications. The company has been located in Washington, D.C., since 1981 and has a nonprofit wing, The Anacostia Energy Alliance, which has contracts with the United States Department of Energy and the Department of Housing and Urban Development.

David Hartman is a management consultant for the Energy and Environmental Division of Booz, Allen and Hamilton, located in Bethesda, Maryland. He has written or cowritten many books and articles on energy conversion.

Harris Mitchell, special consultant for Canada, has worked in the field of home repair and improvement since 1950. He writes a syndicated newspaper column, "You Wanted to Know," and is author of a number of books on home improvement.

John Spears works as a solar systems engineer for Vitro Laboratories in Silver Spring, Maryland. He has been engaged in design engineering and consulting for solar-energy projects since 1972, has constructed a number of solar houses and greenhouses, and is author of more than 100 technical publications on solar energy.

T.E.A., Inc. and T.E.A. Foundation, Inc. are nationally known New Hampshire firms that provide design, engineering, and consulting services for solar heating of residential and commercial buildings. T.E.A. Foundation has produced a series of plans and manuals, titled "Easy To Build Solar," for low-cost solar additions, including thermosiphoning air panels.

Library of Congress Cataloguing in Publication Data
Main entry under title:
  Energy alternatives.
  (Home repair and improvement)
  1. Dwellings—Energy conservation. 2. Renewable energy sources. I. Time-Life Books II. Series.
TJ163.5.D86E515 1982      696      82-10332
ISBN 0-8094-3494-6 (retail ed.)
ISBN 0-8094-3495-4 (lib. bdg.)
ISBN 0-8094-3496-2 (reg. bdg.)

# Contents

# 1 The Energy-efficient Home

Keeping a home warm in winter, cool in summer and supplied with hot water around the clock in all seasons costs a lot of money. In recent years, the prices of electricity, natural gas and fuel oil for home heating and cooling have skyrocketed. Energy consumers today are worrying not only about inroads on the family budget but also about the continued availability of gas and oil—at any price.

In response to these problems, scientists and technologists have redoubled their efforts to find new sources of energy and to make present sources more economical. News reports continually trumpet breakthroughs and near-breakthroughs. While many people are confident that science and industry can assure cheap, dependable energy sources for the future, more and more homeowners are seeking individual solutions to the so-called energy crisis and, in many instances, are increasing the values of their homes at the same time.

For some homeowners, the solutions involve new uses for old and neglected resources. Individuals are fashioning backyard generators that harness wind or water, and they are adding economical wood-burning furnaces to their home heating systems *(Chapter 3)*. Not only do these adaptations substitute cheaper—or free—energy sources for more costly ones, they provide a flexibility that could save the day should standard fuels suddenly become unavailable.

Other homeowners are taking fuller advantage of the greatest of all energy sources—sunlight. Although the earth receives only one two-billionth of the sun's energy, just a quarter hour of sunshine equals all of the energy consumed by humans from all other sources in an entire year. This bounty, trapped in rooftop collectors or glazed, wall-mounted panels, can be readily applied to space and water heating, and it may someday be a major source of electricity.

For most people, however, a wise, personal energy policy starts with putting what they already have to the best use. Any house can be made more comfortable in summer by taking advantage of cool breezes, and the addition of an inexpensive device made of wood and plastic, called a solar chimney, can circulate air throughout a house even on still days *(pages 12-15)*. In winter, shades and shutters that provide effective thermal barriers at windows, air locks that keep cold breezes out of doors and thick walls that contain massive amounts of insulation can cut heating costs dramatically.

In fact the spectrum of choices facing today's energy consumer is sometimes bewildering. Winnowing out the right answer to your energy needs will require patience and perseverance. And first, before examining any of the options, you must undertake a gimlet-eyed appraisal of your present energy supply and the efficiency with which you use it *(overleaf)*.

# Exploring Your Options for Energy Savings

Reduced to its simplest economic terms, the energy you use to run household appliances and to maintain a comfortable climate in your home is a commodity. Whether it comes in the form of gallons of fuel oil, kilowatt-hours of electricity or cubic feet of gas, you buy energy from a utility or fuel company—and probably pay a lot for it.

There are two basic ways to reduce the amount of energy you buy. The first is to employ conservation measures. The second is to harness an alternative source. To discover the ways you can best approach these goals, take stock of your needs and the potential of the resources available. For example, if you have a large family and spend a high percentage of your energy dollars on hot water, your highest priority may be finding a way to save on water heating. If you live in the sunny Southwest, a solar water heating system could cut your fuel bills dramatically; if your home is in a cloudier climate, a heat-pump water heater might be a more practical solution.

The first step in making such decisions is to conduct an energy audit of your house. Begin by analyzing your fuel and utility bills, as explained in the box opposite, to determine how much energy you are using and what you are using it for. Most consumers spend more than half their energy dollars heating or cooling their homes. Many are paying in large part for wasted energy—wintertime heat that escapes through poorly insulated walls and out cracks around doors and windows, or summertime air conditioning that compensates for outside heat creeping in via the same pathways.

To determine whether your heating or cooling bills are excessive for your area, give your house an efficiency rating by means of the formula outlined in the box. To do the arithmetic, you will need to convert your present heating energy requirements into BTUs and to find out your area's annual number of heating-degree days (HDDs)—a measurement used by weather experts to define the severity of a given climate.

A BTU—for British Thermal Unit—is a standard measurement of heat energy. One BTU is the amount of heat needed to raise the temperature of 1 pound of water 1° F. A gallon of fuel oil produces 140,000 BTUs; a kilowatt-hour of electricity 3,400 BTUs; and 100 cubic feet (1 CCF or therm) of gas 100,000 BTUs.

Heating-degree days are calculated by subtracting the average temperature for each day from a base temperature of 65°. For example, a day with an average temperature of 31° is rated at 34 HDDs.

Annual HDD totals range from 1,500 in regions along the Gulf of Mexico to 10,000 along the U.S.-Canadian border. To find the HDD figure for your area, ask your local utility company. Many utilities also calculate cooling-degree days (CDDs), the difference between 65° and the average outdoor temperature. Annual CDD totals range from approximately 200 in the Northwest to more than 3,000 in the Southwest.

The result of your efficiency calculations will be the number of BTUs your house consumes, per square foot, per heating- or cooling-degree day, in order to maintain a comfortable temperature.

If you come up with a poor to average rating, examine the house to see how energy is going to waste. Although you can do this on your own, you may want to get advice from your local utility company or an energy audit firm.

To help consumers find ways to conserve energy, federal law requires utilities either to perform walk-through house inspections for a minimal fee or to provide explanatory literature that will enable homeowners to conduct inspections.

You or the utility auditor should check the condition of the caulking and weather stripping around doors and windows, examine the heating and cooling systems to see if they are at peak efficiency, and rate the insulating ability of your walls, roof and windows in terms of R value.

R value is a measurement of a building material's ability to slow the passage of heat: the higher the R value, the more effective the insulation. The wall insulation in most well-built contemporary homes ranges from R-11 to R-19 and ceiling insulation from R-19 to R-38, depending on the climate. If your house is inadequately insulated for your area, your utility company can advise you on how much additional insulation you will need in order to bring it up to standard.

If the results of your energy audit indicate that your home is operating at peak efficiency, you can start to consider some of the modifications—or retrofits, as they are called—described on the following pages. The effectiveness and practical value of any energy retrofit depends on a combination of several—in some cases, all—of the following factors.

□ CLIMATE. The climate of a region, as rated in heating- or cooling-degree days, gives an indication of how practical certain conservation measures might be. For example, a house located in an area with a high number of heating-degree days will save money with super weatherproofing measures such as triple-glazed windows (page 18) or extra-thick insulation (pages 26-31). Homes in warmer climates—with a high number of cooling-degree days—will benefit more from a ventilating system, such as the solar chimney (pages 10-15), that will cut the load on an air conditioner.

The direction of prevailing winter winds in your region may tell you which walls need insulation or which windows need extra glazing. Your weather bureau can provide data on wind direction.

□ HOUSE SIZE AND SHAPE. The total square footage of a house—or a room—often dictates the size of equipment such as solar collectors. The dimensions and shapes of outside walls determine where (or if) you can mount certain solar devices such as thermosiphoning air panels (pages 48-57). To take the measurements, sketch rough floor plans of the rooms, draw elevations of exterior walls, and note down all the dimensions.

□ SOLAR ORIENTATION. In the Northern Hemisphere, the sun moves across the southern sky. During winter it moves in a much lower arc than in summer, so sunlight striking the south wall of a house in December may be blocked by roof eaves in July. To get the full benefit of its warming rays, solar collecting devices must be mounted to face within 20° of true south—the more directly, the better. The sun lies true south at noon.

True south, also called due south, differs from the magnetic south shown on compasses because of an anomaly that is known as the magnetic deviation. The discrepancy between true and magnetic south varies from one area to another, but any local land surveyor can tell you what it is at your site. To locate true south on your own property, either check a survey map for your house lot or take a compass reading to find magnetic south, then compensate by the amount of the magnetic deviation in your locale.

□ LATITUDE. The distance of an area from the equator is expressed in degrees of latitude—a figure available in any atlas. To catch noon sunlight face-on, solar collectors are usually mounted tilting at an angle equal to or within 10° of the latitude in which they are located.

□ INSOLATION. Insolation is the amount of sunlight your area receives. It is expressed in terms of BTUs per square foot or in langleys. A langley equals 3.69 BTUs. Insolation is a critical factor in calculating the potential output of solar devices and, along with the dimensions of the house, is one of the statistics used in determining how many square feet of collector surface you need. High rates of insolation will also indicate the need for heavy summertime window shading. Some utility companies and weather services record local insolation data. If you cannot get the information, write or call the National Climatic Center, Federal Building, Asheville, North Carolina, 28801.

□ POTENTIAL OUTPUT. Some energy devices, such as heat pumps and wood furnaces, are rated according to how many BTUs per hour they are capable of producing—information that the manufacturer or dealer can furnish. To estimate how well a proposed installation will handle your needs, weigh its BTU output against your BTU requirements, as figured in the energy audit. Solar collectors are rated according to how many BTUs they can deliver per square foot of glazing area during daylight hours. A solar-equipment distributor may have performance statistics for sample climates to help you size your collector.

□ COST EFFECTIVENESS. The cost effectiveness of a retrofit is calculated by making an estimate of its payback period—the time it will take for the installation to pay for itself. For example, a $600 heating device that can handle 10 per cent of a home's space-heating needs will trim $100 a year from a $1,000 annual oil heating bill. It will pay back its costs in six years. (When figuring costs of a retrofit, check to see if it qualifies for tax credits, which will help defray its expense.)

Not all retrofits can be counted on for quick payback periods. But favorable payback is only one reward for installing energy retrofits. An installation that results in noticeable savings can add to the resale value of the home. For many owners, there is a value that cannot be measured in dollars and cents: the personal satisfaction of conserving energy.

## Scrutinizing Your Costs

Begin an energy audit by obtaining records of the monthly bills for the past three years from all the utility or fuel companies from whom you buy energy. The records should list the number of units of fuel charged to the house and the total price per billing period. Do all of the calculations in terms of units of fuel. You can later convert the figures to dollars by multiplying by the unit price.

To find out how much energy your water heater uses, subtract the ground-water temperature from the hot-water temperature. Multiply that figure times 20 (the gallons of hot water a typical person uses each day) and that by 8.33 (the number of pounds in a gallon). Then multiply the result by the number of people in your household and by 30 to get the total number of BTUs required for water heating each month. For gas or oil water heaters, factor in an efficiency rating by dividing the total by .75 if the heater is less than 10 years old or by .5 if it is more than 10 years old. Convert the BTUs to units of fuel.

To isolate air-conditioning and heating fuel consumption, first calculate your home's base load—the more or less fixed monthly energy needs of lights, appliances and water heater. Then subtract the base load from a month's total utility bill.

To find the base load for an all-electric house without air conditioning, average the bills for June, July and August by adding up the kilowatt-hours consumed and dividing by three. For an all-electric house that has air conditioning, determine the base load by finding two months in a year when the house was neither heated nor cooled; typically, electricity use is lowest during May and September. Average the kilowatt-hours used in these two months and subtract the result from your winter bills to isolate the heating needs. To determine a month's air-conditioning costs, subtract the base load from a summer month's electricity bill.

For a house with an oil furnace and water heater, but electric appliances and air conditioning, average the summer oil bills to get the base load for oil and subtract the result from the winter bills to get heating consumption. To calculate air-conditioning needs, subtract the average winter electricity use from the summer months' bills.

For a house that burns natural gas for heating, hot water and major appliances, average the summer months' gas bills to find the base load for gas usage, and subtract it from the winter gas bills to isolate your heating needs.

After you have figured the amount of fuel you use to heat or cool your house, determine whether your consumption is excessive. First, convert the units of fuel bought each month into BTUs: Multiply kilowatt-hours by 3,400; gallons of oil by 140,000; and therms or CCF (100 cubic feet) of gas by 100,000.

Next, divide the BTU total by your local heating or cooling degree-day rating for that year. Convert this number to a square-footage basis by dividing it by the living area of your house, and you get a rating of the house's heating or cooling efficiency. For heating, a rating up to 10 is excellent; from 11 to 20, average; more than 20, poor. For cooling, a rating up to 3 is excellent; 3 to 6, average; more than 6, poor.

Thus, a 2,000-square-foot house that used 1,400 therms of gas per year with an HDD rating of 5,500 HDDs would be rated at 12.7, in the average range (1,400 x 100,000 = 140,000,000 ÷ 5,500 = 25,455 ÷ 2,000 = 12.7).

# Cooling the House Grandma's Way

Before the advent of electric air conditioners, homeowners used an array of simple tactics to beat the heat. They pulled the shades on the sunny side of a house (but left the upper sash of double-hung windows open), then opened both the shades and windows on the shady side, and turned on fans that stirred cooling breezes.

These battlers of the heat had an instinctive appreciation of at least three complex cooling processes. In the jargon of today's energy pioneers—whose interest in oldtime cooling methods has been sparked by the quest for lower utility bills—the processes would be identified as preventing solar gain (pulling the shades), facilitating personal evaporative cooling (letting breezes cool your sweat-dampened skin), and utilizing thermosiphoning and active ventilation to augment air flow (lowering upper window sashes, providing vents in the attic and turning on fans).

Of these processes, only thermosiphoning is difficult to understand.

Thermosiphoning is based on the principle that heat rises. If windows near the ceiling of a room—or vents near the top of the house—are open, the rising hot air will waft outside. As this air leaves the house, it will be replaced by cooler air sucked into open windows on the shady side of the house or crawl-space vents near the cool ground. To take fullest advantage of thermosiphoning, open the lower sashes of windows on the cool side of a room about 25 per cent less than the upper sashes of the windows on the warm side. The restricted passage on the cool side creates a natural pressure that actually pulls the air in faster, increasing evaporative cooling.

On a smaller scale, at individual windows, thermosiphoning sometimes has a negative effect that must be avoided. Between loose-fitting draperies and the inside of a window is a space that collects hot air on sunny days. When this air rises, it pulls cooler air up under the fabric, creating a convective current that circulates the hot air into the room. A valence of fabric or wood installed across the top of the window will reduce this flow.

On a scale larger than the treatment of individual windows, thermosiphoning can be used to encourage air flow throughout an entire house. Gable and soffit vents have long been used to flush unwanted heat out of attics, and glazed cupolas—especially popular atop homes in New England—provide a similar siphoning service for the highest living space of a house. Today many homeowners are duplicating the effect of the cupola with a plywood and plastic shaft, called a solar chimney.

In a solar chimney, sunrays heat the air through glazed southern and western faces; the heated air rises quickly out of the chimney top, pulling heated air out of the house in its wake. A wind-driven turbine, capping the chimney at the top, can further increase the air flow when the wind is blowing. As the heated air exits from the top of the house, cooler air is pulled in through floor registers from the basement or crawl space, and through windows opened to shaded porches or gardens.

For maximum efficiency, solar chimneys should be positioned on an unshaded expanse of roof that receives full midday and afternoon sun. Choose a portion of the roof offering a straight descent through the attic to a room below, free of intervening beams, heating ducts or electrical cables. If necessary, remove a section of attic flooring to make an inspection before proceeding.

For the body of the chimney, you will need standard, construction-grade, 2-by-4 studs and ½-inch exterior plywood. You can use the same plywood to fashion sliding doors for the ceiling below, which are used to close off the opening in cooler weather. Transparent, fiber-reinforced plastic panels ¼ inch thick, sold under the trade name Fiberglas, are used to glaze two sides of the upper portion of the chimney. A standard 16-inch wind-turbine vent is set atop the chimney. All of these materials are available at large building-supply stores.

The turbine requires a square chimney cap, however, which is not standard. Take the turbine to a sheet-metal worker to have a cap built to the dimensions of your chimney—15½ inches square for houses that have rafters or roof trusses spaced 16 inches apart, 23½ inches square for roofs based on 24-inch spacing. The cap should have a down-turned lip around its perimeter and an upturned collar at its center to accommodate the turbine. To prevent water from leaking through the joints of the chimney, you also will need lengths of aluminum flashing, 4, 8 and 18 inches wide.

Silicone caulking and asphalt roofing cement finish the job outside, sealing all the rooftop seams. To seal the inside, you need two lengths of U-shaped aluminum channel, each 1½ inches wide and twice the length of the ceiling opening below the chimney (page 15, Step 1). Also buy rigid foam insulation, 1 inch thick, to glue to the back of the sliding doors.

As with any building project, check with your local building department before you begin work; in many communities, local building or architectural codes strictly regulate the design and position of rooftop additions. In addition, you may have to obtain a building permit.

## New Look at an Old Friend

A tried and true ventilator, the electric, paddle-style ceiling fan has been in use in warm climates for more than 100 years. Today it is gaining renewed favor in areas where the cost of electric air conditioning has sky-rocketed. In warm weather, the fan circulates standing room air, pulling cool air from the floor of a room and washing it over the room's occupants. In winter, the same fan can force warm air that has risen back down into the living space.

A ceiling fan may be equipped with adjustable paddles; reversing the pitch of the blades reverses the flow of air. More convenient are models that have a reversible motor, which spins the blades in one direction to force the air flow up, in reverse to push it down. Additionally, most fans have variable-speed motors that let you slow the fan for gentler, draft-free downward circulation of warm air, or speed it up for a

brisker, high-speed current for cooling.

Paddle fans can be mounted on any flat or gently angled ceiling that is at least 7 feet high. Most fans are installed in the same way as a hanging pendant light fixture, hooking over a metal strip

on the bottom of an electrical ceiling box. For heavier models, a box with a threaded stud must be installed. Most manufacturers provide instructions for installation and hardware matched to each model of fan they sell.

## Making the Most of Air Currents

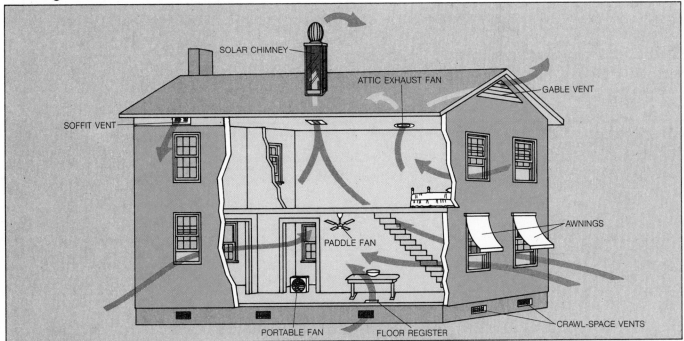

SOLAR CHIMNEY

ATTIC EXHAUST FAN

GABLE VENT

SOFFIT VENT

AWNINGS

PADDLE FAN

PORTABLE FAN

FLOOR REGISTER

CRAWL-SPACE VENTS

**An air-cooled house.** Cool air enters this house through crawl-space vents, passing through floor registers into the living spaces. Windows, open at the top on one side of the house and at the bottom on the other, augment the natural crosscurrent with a thermally siphoned flow. Windows exposed to the south are shaded with ex-

terior awnings on the first floor to keep out the sunlight; windows on the upper floor are shaded by the roof overhang.

Increasing the air flow through the whole house are an attic exhaust fan, mounted in the ceiling of an upper-story room, and a solar chimney,

which expels heated air out of the house from the highest point. Air circulation through the attic is provided by louvered gable vents at each end of the house and soffit vents underneath the eaves. A paddle-style ceiling fan over the dining area and strategically placed portable fans create additional cooling breezes.

# Building a Solar Chimney

**1** **Cutting the roof opening.** Inside the attic, drill a locating hole through the roof next to a rafter to mark the position of one upper corner of the chimney. Measure the distance between the rafters. On the roof, find the locating hole, then use a linoleum knife to remove shingles from an area 2 inches larger than the planned chimney. From the locating hole, draw a cut line parallel to the roof ridge and as long as the distance between rafters.

Place the chimney cap with its upper edge on the cut line and level it with scraps of wood. Hold a carpenter's level plumb at the corners of the cap closest to the eaves and mark their locations on the roof. Make a second mark ¾ inch up the slope from each of these corner marks. Draw a straight line between the new marks and between each end of this line and the first cut line. Cut along all four lines with a saber saw.

Hang a plumb bob from each corner of the roof opening. Have a helper mark where the bob touches the joists below and cut away the insulation between the marks.

CHIMNEY CAP

LOCATING HOLE

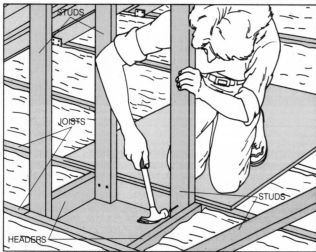

STUDS

JOISTS

STUDS

HEADERS

**2** **Framing and enclosing the inside shaft.** Nail two short 2-by-4 boards, called headers, between the joists just outside the marks. Cut four 2-by-4 studs long enough to reach from between the joists to a point at least 1 foot above the level of the roof ridge. At the intersections of the headers and joists, nail the studs to the joists. Then nail the studs to the rafters.

If the rafters and joists are offset, and not one directly above the other as in the truss roof shown here, you will have to use shims and blocking at the joists to keep the studs plumb.

Drill four locating holes down through the ceiling at the corners formed by the studs and headers. Finally, nail plywood to the outsides of the studs to enclose the chimney within the attic.

SPREADERS

SPREADERS

**3** **Building the rooftop section.** To make supporting frames for the plywood and the plastic glazing, nail four short 2-by-4 spreaders horizontally between the studs at the top. Add a second set—also horizontal—where the chimney passes through the roof. Enclose the entire north and east sides of the chimney with ½-inch exterior-grade plywood; then, using the same kind of plywood, enclose the lower portion of the south and east sides up to the middle of the bottom spreaders. Note: Use only aluminum or galvanized nails any place where the nailheads will be exposed to the weather. Paint all of the interior surfaces of the studs, spreaders and plywood with flat black, oil-base paint.

**4** **Glazing the sunny sides.** Cut two sheets of plastic glazing to the exact size of the south and west sides of the chimney. Drill screw holes ³/₁₆ inch in diameter at each corner of the sheets and at 12-inch intervals around the perimeters. After mounting the roof, brace the glazing against the chimney frame and drill ¹/₁₆-inch pilot holes into the wood through each hole in the plastic. Set the plastic aside and apply a thick bead of silicone caulking to the frame. Return the plastic to the frame and secure it with round-head wood screws.

To seal the corners, cut four lengths of 4-inch-wide aluminum flashing to the height of the chimney at each corner. Bend each piece in the middle lengthwise and attach it to the corners of the chimney with round-head aluminum screws spaced every 6 inches.

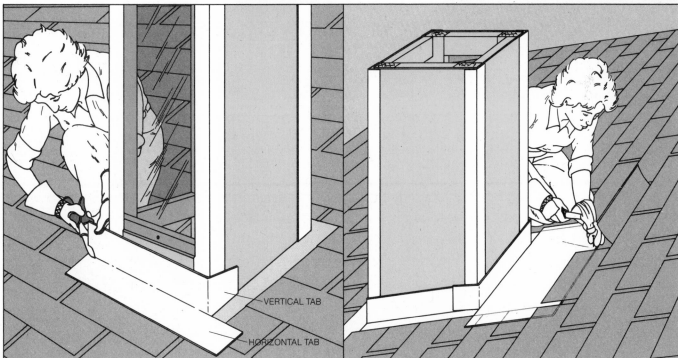

VERTICAL TAB

HORIZONTAL TAB

**5** **Flashing around the chimney.** Cut an 8-inch strip of flashing 8 inches longer than the width of the side of the chimney closest to the eaves. Bend the strip in the middle lengthwise. Place the strip at the joint and cut 4 inches into the bend at each end with tin snips (above, left). Bend the vertical tabs created by the cuts sharply around the corners of the chimney; insert the horizontal tabs under the shingles at the ends. Secure the flashing to the roof with roofing nails. Do

not nail the flashing to the chimney. Similarly cut and emplace flashing along both of the slanting sides, inserting the horizontal edge of the metal under the shingles at the sides.

Remove two rows of shingles above the remaining side of the chimney. Cut an 18-inch strip of aluminum 12 inches longer than the width of the chimney. Bend the strip lengthwise, 4 inches from one edge. Make 6-inch cuts into each end

along the bend and install the strip with the 4-inch portion vertically against the chimney. Bend the tabs around the chimney as before, and insert the horizontal edge of the metal under the shingles above the joint. Carefully bend back the shingles to nail the metal directly to the roof (above, right). Replace one row of shingles, inserting and nailing them under the shingles above them. Seal all of the nailheads and seams with roofing cement.

13

**6** **Completing the flashing.** Cut lengths of 6-inch flashing for the four sides of the chimney, each long enough to extend 6 inches around the corners *(inset)*. At the side closest to the eaves, hold the flashing against the chimney with the bottom edge of the metal ¼ inch above the flashing installed in Step 5 *(page 13)*. Bend the ends around the corners and fasten the metal to the chimney with roofing nails. In the same way, attach flashing to the sloping sides. Then install flashing on the uphill side of the chimney, bending its ends to overlap the side pieces. Seal the nailheads and joints between the metal and the chimney with roofing cement.

EAVES

**7** **Installing the turbine vent.** Fit the chimney cap onto the chimney top and nail it to the two plywood sides. On the remaining sides, drill through the metal and the plastic, and secure the cap with screws. Fit the round base of the turbine vent onto the collar of the chimney cap, joining the two with sheet-metal screws. Finally, rotate the adjustable rings at the base of the vent to straighten the turbine.

Paint the wooden sections of the chimney to match the exterior trim of the house; then, after allowing at least 48 hours for the paint to dry, seal the seams at the base of the chimney cap with silicone caulk.

COLLAR

ADJUSTABLE RINGS

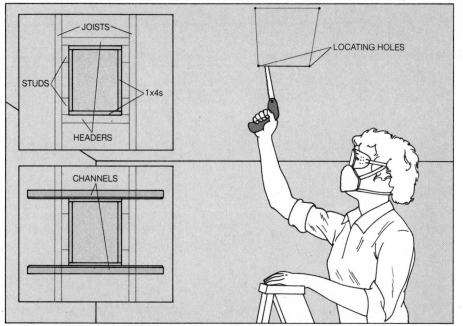

## Adding the Access Doors

**1 Cutting the opening.** Connect the four locating holes drilled in Step 2 *(page 12)* to outline the ceiling opening and, wearing safety goggles and dust mask, cut along the lines with a keyhole saw. Join four 1-by-4s to make a frame that will just fit into the opening, and secure the frame to the faces of the headers and studs with finishing nails *(inset, top left)*.

With a hacksaw, cut 1½-inch aluminum U channelling—so called because of its U-shaped cross section—to twice the length of the opening. Drill ³⁄₁₆-inch holes every 6 inches along the side of each channel and screw the channels through the ceiling into the headers. The open sides of the U should face each other across the ceiling opening *(inset, bottom left)*.

**2 Inserting the doors.** Cut a plywood rectangle 1 inch larger than the size of the ceiling opening. Glue rigid foam insulation onto the back of the plywood. Cut the panel exactly in half with a circular saw. Bore ¾-inch holes in both pieces at each end and insert finger pulls. Push channel caps into the channels at one end and slide both door sections into the channels from the other end. Add a second set of channel caps to keep the doors from sliding out.

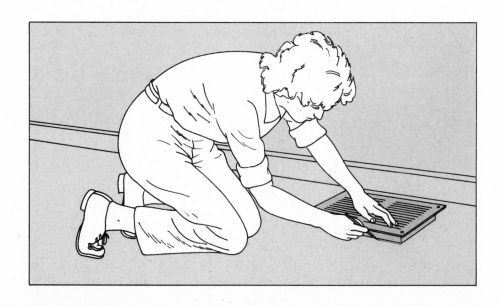

## Tapping the Coolness of a Crawl Space

**Adding a floor register.** From the crawl space or basement, find an area clear of any electric wires or plumbing pipes that would interfere with the opening for the new register. Drill a locating hole straight up through the floor next to a joist in the clear area. In the room above, put the register on the floor with one corner on the locating hole. Trace around the base collar of the register with a pencil, then cut out the outlined area with a saber saw.

Set the register into the opening and push it down until the cushioned flange underlying the grate lies flat on the floor. Secure the register with wood screws at each corner.

# Window Treatments That Save Energy

Windows are notorious energy wasters. In summer they allow sunlight into a house, overheating it and driving up air-conditioning costs. In winter they let precious heat escape, sending fuel bills soaring. Some 25 to 30 per cent of a typical home's heat goes out the window during cold months; in older houses, with loose-fitting windows, the loss can exceed 50 per cent.

Windows act as heat passageways because glass is a very poor insulator. Most windows have an insulation rating, or R value (pages 8-9), of less than 1.

Fortunately, keeping outside heat from coming in through a window is easy, because much of it enters in the form of sunlight. You need only cover the window glass or direct the sunlight away from it. A variety of conventional shades, blinds, shutters, and awnings all effectively block out the sun. A louvered overhang, like the one on page 18, built with preservative-treated wood, will bounce heat-producing rays away from the glass while still allowing some indirect light to filter through.

Controlling heat loss through windows is more difficult, in part because heat not only exits through the glass but leaks out through cracks around the frame. Caulking and weather-stripping window frames will dramatically reduce the outflow of heat. But the most effective solution is to insulate the windows themselves with permanent or seasonal layers of glazing, and then cover them on cold nights with insulated shades or shutters.

No glazing material is in itself a good insulator, but a window's R value can be doubled by adding a second layer of glazing, which creates an insulating pocket of trapped air. The insulation can be increased by another third, to almost R-3, by adding a third layer of glazing.

One of the simplest and most commonly used methods of adding glazing is to install exterior storm windows fitted with single or double panes of glass or rigid plastic. A less expensive but equally effective technique is to fashion interior windows of rigid plastic or plastic film. The plastic is either mounted on frames (available in kits) or secured directly to the window casing with screws or tape.

A third technique for increasing layers of window glass is to replace individual panes with sealed, double-glazed panes, ⅝ inch thick. If the window sashes cannot be altered or reconstructed to accommodate the thicker panes, the entire window can be replaced by a prefabricated unit with double or triple glazing.

When adding multiple glazing, keep in mind that the south windows of a house serve as solar collectors in cold weather, transmitting warming, energy-conserving sunshine into the home. A second layer of glazing added to these windows should be one of the high-transmittance materials listed in the chart opposite. Ordinary window glass is transparent, but it has a shiny, reflective surface that allows sunlight to bounce off. The somewhat duller-looking high-transmittance materials absorb most of the sun's rays, allowing more heat to pass through.

Usually it is not worthwhile to triple-glaze south windows. Whatever material is used, a third layer will reduce the passage of light, and the subsequent loss of solar heat in cold climates may actually offset the heat-containing value of the added insulation.

To reduce heat loss further, windows may be blocked out completely at night, using a variety of opaque insulating materials. Lightweight fabric insulation, available at many fabric and shade stores, can raise the insulating value of a single-glazed window to R-6. On a multi-glazed window, the extra insulation can raise the total R value close to that of the surrounding walls.

The insulation, which comes in large, blanket-like sheets, generally consists of two layers of dense polyester fiber—similar to the material used to fill bedroom quilts and outdoor clothing—sandwiching a plastic-film vapor barrier. The insulation is faced with moisture-resistant fabric; some brands also include a heat-reflective metallic shield. Covered with decorative drapery fabric, the thick insulating material can be fashioned into a variety of heat-saving window coverings; the accordion-pleated Roman shade on pages 19-21 is one example.

Flexible magnetic strips backed with adhesive—available at many hardware and fabric stores—are attached to the edges of the shade. Pressed against a steel window frame—or against matching magnetic or steel strips mounted on a wooden frame—the magnets seal the window against leaks.

Interior shutters of rigid foam insulation are an inexpensive alternative to bulky fabric window coverings. Pop-in shutters (page 22) of ordinary styrene foam board—similar to the material used for inexpensive picnic coolers—have an R value of 4 to 5 per inch of thickness. Even better are shutters made from urethane foam or foil-sided polyisocyanurate foam—rated at R-8 per inch and used to insulate exterior walls under such trade names as Thermax and R Max.

Pop-in shutters have several drawbacks: They must be removed and stored during the daylight hours; they tend to shrink or warp with age; and unless coated with fire-retardant paint, they may burn or give off toxic fumes when exposed to flame. Hinged bifold shutters (pages 22-23), faced with flame-resistant fabric or a thin skin of wood or plastic, take longer to construct; but once installed, they are sturdier, less hazardous and far more convenient.

Dozens of other window treatments are available in kit or manufactured form. Those pictured on pages 24-25 are a representative sampling of ingenious space-age products that can transform energy-wasting windows into energy-conserving home accessories.

# The Basic Glazing Materials

| Material | Common trade names | Form | Characteristics |
|---|---|---|---|
| Window glass | | Rigid single sheets $\frac{1}{16}''$ to $\frac{3}{16}''$ thick; double-strength ($\frac{1}{8}''$) is standard and can be cut to any size or shape. Hermetically sealed, double-wall sheets $\frac{7}{16}''$ thick are available in standard window sizes. Both types are available in clear, tinted, frosted or reflective sheets. | Most durable and readily available glazing material. Heavier and requires stronger framing than plastics. |
| Low-iron glass | Sunadex<br>Solakleer<br>Heliolite | Rigid single sheets $\frac{1}{8}''$ thick, or double-wall sheets $\frac{7}{16}''$ thick; precut in most standard window sizes. | Reduced iron content and nonreflective surface increase light transmittance, making it the best glass for storm windows. |
| Acrylic | Plexiglas<br>Lucite<br>Acrylite<br>Exolite | Rigid single sheets $\frac{1}{8}''$ to $\frac{1}{4}''$ thick, double-wall sheets $\frac{5}{8}''$ thick; cut by dealers to size from 4'-by-8' panels. | Excellent light transmittance, durability and light weight make it a good—but expensive—glazing material. Semitransparent, double-wall sheets are used for privacy in large windows. Scratches easily. |
| Polycarbonate | Tuffak<br>Lexan<br>Exolite | Rigid single sheets $\frac{1}{8}''$ to $\frac{1}{4}''$ thick, double-wall sheets $\frac{1}{4}''$ to $\frac{5}{8}''$ thick; similar to acrylic and cut in the same manner. | More resistant to impact and high temperature than acrylic, but transmits less light. Discolors with age. Semitransparent, double-wall sheets are virtually indestructible, thus useful for home security. |
| Fiberglass-reinforced polyester (FRP) | Filon<br>Sun-Lite<br>Crystalite<br>Glasteel | Flexible sheets in flat, corrugated or shiplap configurations, .025'' to .060'' thick, in widths to 60'' and lengths to 50'; also available in more rigid multiple-layer sheets. | Translucent but not transparent. Strong, light and inexpensive. |
| Acrylic-polyester laminate | Flexigard | Two-ply film 7 mils thick, in rolls 4' wide, 20' to 150' long. | Translucent; combines acrylic durability with polyester flexibility. Should be installed with acrylic side facing out. |
| Polyester | Mylar<br>Llumar<br>Sun Gain<br>Heat Mirror | Film 1 to 7 mils thick, in rolls 26'' to 60'' wide, 50' to 300' long. Available in a variety of tints or coated with metallic, reflective surfaces. | Inexpensive, high-transmittance material for interior use when treated with ultraviolet stabilizers; untreated film deteriorates quickly. Tears and wrinkles easily. |
| Polyethylene | | Film, usually manufactured in the thicknesses of 4 or 6 mils; available in rolls 10'' to 42'' wide, up to 150' long. | Least expensive, but degrades rapidly and melts in moderate heat. Adequate for temporary use. |

**Choosing a glazing.** This chart lists, by both generic and common trade names, the glazings best suited for insulating windows—from traditional glass to film as thin as a plastic sandwich bag. All of the glazing materials are either transparent or translucent, and many of them are available in a variety of tints or with decorative finishes and coatings. All of the rigid materials are available in what is known as double-wall form: prefabricated double-glazed sheets, with two layers that are connected and held parallel by sealed edges or interior connectors.

# Multiple-pane Windows for Comfort and Convenience

**Adding layers of glass and plastic.** Three techniques for adding layers of glazing are illustrated here, as seen from the outside of a double-hung window. Double glazing, the most effective and most expensive of the methods, involves replacing single-pane glass with double-wall insulating glass—two ⅛-inch panes separated by a ⅜-inch air space. Panes with triple glazing are also available.

The combination storm window and screen is somewhat less efficient than double-glazed window panes or weather-stripped single-pane exterior storm windows; air leaks through the cracks around the panels. But the combination window is convenient: It never has to be removed, and in summer it can be quickly transformed into a screen for open-window ventilation. Thin plastic film mounted inside the window is an inexpensive alternative or addition to any of the other installations and works equally well.

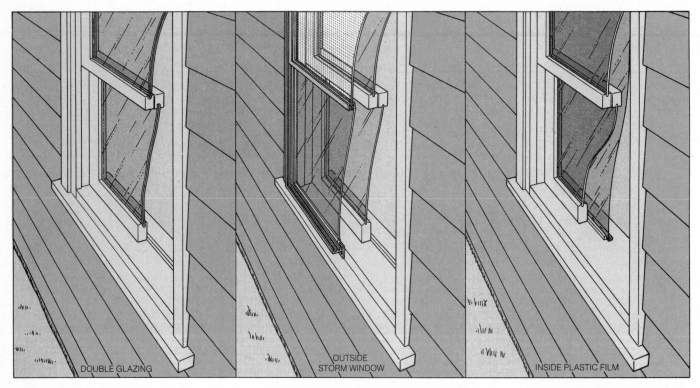

DOUBLE GLAZING

OUTSIDE STORM WINDOW

INSIDE PLASTIC FILM

## A Louvered Overhang to Deflect the Sun

**Making the overhang.** Using 2-by-8s, cut two side supports for the louvers, each 3 to 4 feet long, and front fascia and rear mounting boards the width of the overhang. Mark louver positions every 8 inches at 60° angles on each support. For louver slots, rout ½-inch grooves 6 inches long and 1 inch wide, or nail 1-by-1 cleats to the supports (*top inset, right*). Nail the 2-by-8s together with 3½-inch galvanized nails.

To mount the box (shown here above a sliding door), screw lag bolts through the mounting board into the header. Temporarily prop the box with long boards. Nail two 2-by-4s vertically to the house wall as nailing surfaces, and cut two additional 2-by-4 braces at opposite 45° angles, as shown. Nail them to the outer edges of the nailers and the underside of the box. Cut louvers from 1-by-6 lumber, and slide them into the louver slots (*bottom inset, right*).

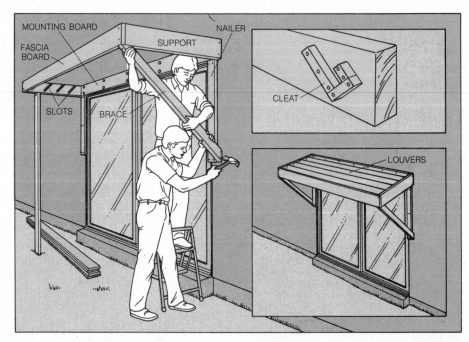

MOUNTING BOARD

FASCIA BOARD

SUPPORT

NAILER

SLOTS

BRACE

CLEAT

LOUVERS

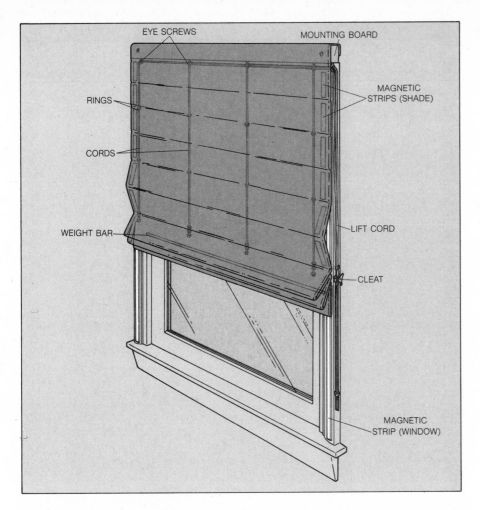

EYE SCREWS

MOUNTING BOARD

RINGS

MAGNETIC
STRIPS (SHADE)

CORDS

WEIGHT BAR

LIFT CORD

CLEAT

MAGNETIC
STRIP (WINDOW)

## Blanketing Windows with Insulated Fabric

**Anatomy of a Roman shade.** An accordion-pleated Roman shade is raised and lowered by vertical cords threaded through rings tied to the fabric and through eye screws set into the bottom of a mounting board. The cords are gathered into a single, knotted lift cord at the side of the window. When the lift cord is pulled, the shade gathers up into 4-inch folds. To keep the shade raised, the lift cord is secured to a cleat on the window casing or on the adjoining wall.

Lowered, the shade seals the window tightly. Flexible magnetic strips sewed into the fabric edges are attracted either to a steel window frame or, in the case of wood-frame windows, to matching magnetic strips or steel straps mounted on the molding. A weighted metal bar holds the bottom of the shade snug against the sill.

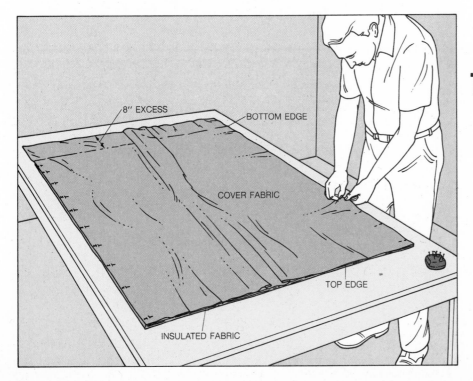

8" EXCESS

BOTTOM EDGE

COVER FABRIC

TOP EDGE

INSULATED FABRIC

## A Roman Shade with a Magnetic Seal

**1** **Preparing the fabric.** Cut a piece of insulated fabric 4 inches longer and 2 inches wider than the dimensions of the window. (For wide windows, you may need to stitch two lengths of insulated fabric side to side.) Cut a length of decorative cover fabric—preferably sheeting or tightly woven drapery material—3 inches wider and 8 inches longer than the insulated fabric.

Lay the insulated fabric with its outside face up on a table, and lay the cover fabric, face down, on top of it. Align the top and side edges of the two fabrics; pin the sides together, allowing the cover fabric's 8 inches of excess length to overhang the bottom of the insulation. Using a sewing machine, stitch a seam ½ inch in from each side of the insulated fabric; then, on each side, stitch a second, zigzag seam as close to the edges as possible, penetrating all layers of insulated fabric. The excess width of the cover fabric will lie loosely on top of the insulation.

**2 Attaching magnetic strips.** Use tailor's chalk to mark fold lines on the sides of the shade, spacing the marks every 4 inches. Cut adhesive-backed magnetic strips into 3½-inch lengths; round off the sharp corners with scissors. Place the magnets outside the straight seam on each side, centering them between chalk marks. Do not put magnets at the top or bottom 4 inches of the seam lines. Fold back 4 inches of the overhang at the shade bottom, then fold another 4 inches and set two more magnets on the front side of the folded cover fabric.

Turn the shade right side out (*top inset, below*), so that all of the magnetic strips are concealed (*bottom inset, below*) except the two at the unfolded bottom; these will eventually be hidden by the hem. Smooth out the shade, ironing it with a cool iron if necessary.

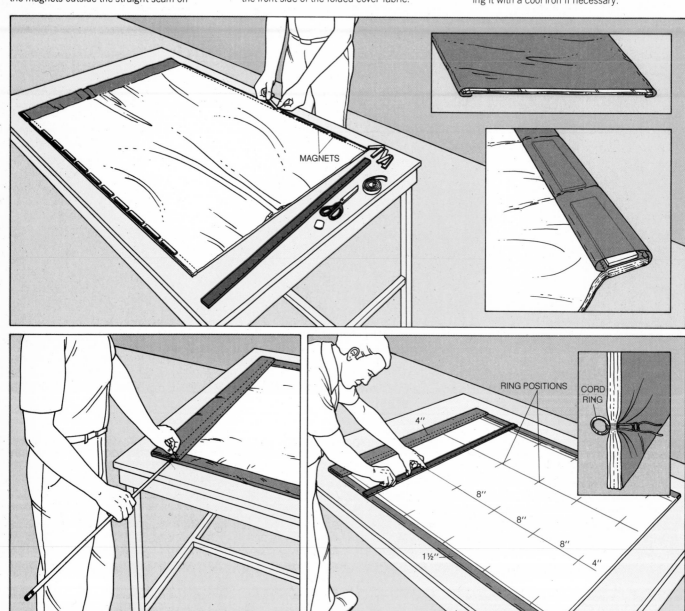

MAGNETS

RING POSITIONS

CORD RING

4''

8''

8''

8''

1½''

4''

**3 Weighting the hem.** With the shade right side out and face down, make a hem by again folding the overhang up 4 inches, then another 4 inches. Machine-stitch the hem along its top edge, then 1 inch below this stitching to form a channel. Cut a ⅜-inch steel rod 2 inches shorter than the width of the shade. Coat the rod with rust-resistant paint or varnish, and tape its ends to protect the fabric. Slide it into the channel, leaving an inch at each end for the bottom magnets; then hand-stitch the sides of the hem.

**4 Attaching the rings.** With tailor's chalk, mark vertical lines on the back of the shade, starting 1½ inches from each edge, spacing the lines at equal intervals of 8 to 12 inches. Mark horizontal lines on the shade every 8 inches, starting 4 inches from the top. (Some insulation is pre-stitched every 8 inches, so measuring is unnecessary.) Where vertical and horizontal lines meet, sew ½-inch cord rings through all layers of the fabric (*inset*), using heavy thread. Tie a square knot, and cut off the excess thread.

**5 Attaching the mounting board.** Cut a 1-by-2 board the width of the shade. Align it with the top of the fabric, and mark its bottom edge to correspond with each vertical row of rings. Insert ½-inch eye screws into the board at each mark, and a final one at the far left end of the board. Drill ¼-inch screw holes through the board 1 inch in from each end. Then wrap the top 4 inches of the shade over the top of the board, and fasten the fabric to the wood with staples or tacks placed 2 to 3 inches apart.

**6 Stringing the cords.** For each vertical row of cord rings, cut a length of nylon shade cord about twice the length of the shade. Tie each cord to the bottom ring, securing the knot with a few drops of white glue. String each cord up through its rings, threading it from right to left through the eye screws above. All of the cords will pass through the eye screw at the far left edge of the shade. Tie the cords together just beyond the last eye screw, and knot them once again several feet down from the first knot. Trim the ends.

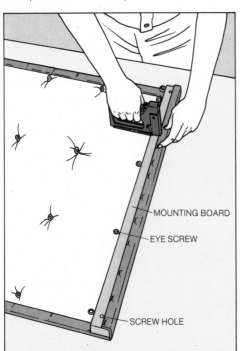

MOUNTING BOARD

EYE SCREW

SCREW HOLE

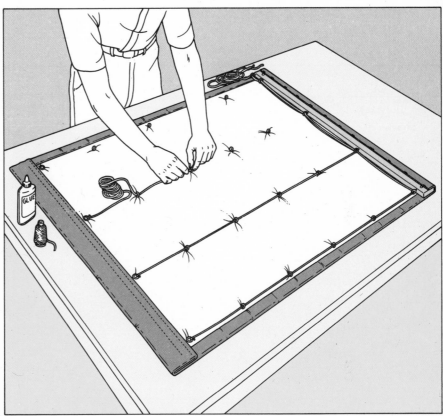

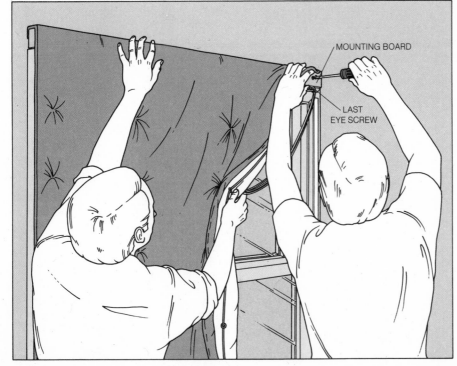

MOUNTING BOARD

LAST EYE SCREW

**7 Mounting the shade.** While a helper holds it at the window, check the shade for squareness and fit; adjust as needed. Uncover the screw holes on the mounting board, and attach the board to the wall above the window, or to the top of the window frame, using 1½-inch screws.

Test the shade by pulling the lift cord firmly downward. The shade should rise evenly in 4-inch folds that bunch loosely below the mounting board. If raising the shade is difficult, replace the last eye screw with a pulley; or attach a pulley to the wall 3 inches beyond the mounting board, then run the lift cord through the pulley. Attach a metal cleat to the wall or window casing to hold the lift cord when the shade is raised.

For a metal-frame window, simply lower the shade and press its sides against the frame; the magnets will hold the shade. With wood-frame windows, clean the frames with isopropyl alcohol, then mount either continuous strips of magnetic tape or thin steel straps directly behind the shade edges. Press the magnets against the strips on the frame to seal the window. Pull the shade out from the bottom to break the seal.

# Inside Shutters for Nighttime Warmth

**Two insulated shutters.** Both of these removable interior shutters cover a window with rigid foam insulation. The edges of the pop-in shutter (*below, left*) are protected by aluminum-foil tape, with adhesive-backed neoprene at the top and bottom edges. Magnetic tape on the window side of the shutter holds it to metal shelf standards or to strapping screwed onto the window casings. Duct-tape handles on the sides of the panel break the magnetic seal for removing the shutter from the window.

The more decorative bifold shutter (*below, right*) is made of two foam boards, covered with wooden door paneling and hinged together. The vertical edges are covered with neoprene weather stripping; the top and bottom edges fit against weather-stripped stops. At night, when the shutter is closed and latched, the weather stripping seals the windows tightly. During the day, the shutter can be folded to one side.

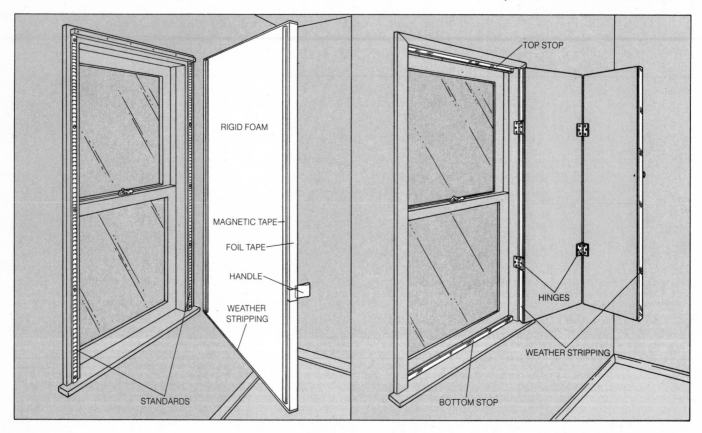

# Building a Bifold Shutter

**1** **Assembling the frames.** After measuring the window, build two four-sided frames for the shutter panels, using 1-by-2 kiln-dried lumber. Allow a ³⁄₁₆-inch clearance between the outer edges of the shutter and the window frame and a ⅛-inch gap between the panels. After nailing the framing pieces together, place the frames in the window; adjust the fit if necessary.

Lay the frames on the back of a sheet of ⅛-inch plywood veneer, and mark the outlines, tracing each frame twice. Cut out, using a circular saw fitted with a plywood-cutting blade. Glue a cut section of veneer onto one side of each frame with carpenter's glue. Secure the veneer to the frame with ⅝-inch brads placed at each corner and every 6 inches along the edges.

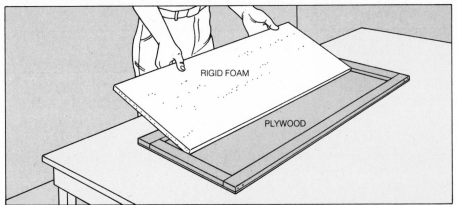

**2** **Insulating the shutter.** Cut two pieces of ¾-inch rigid foam insulation to the inside measurements of the frames, allowing a ¹⁄₁₆-inch clearance on all four sides. Lay the insulation into the frames *(left)*, and enclose the insulation with veneer. Set all the brads in the veneer with a nail set, fill the holes with matching wood putty, and sand the edges with 120-grit sandpaper.

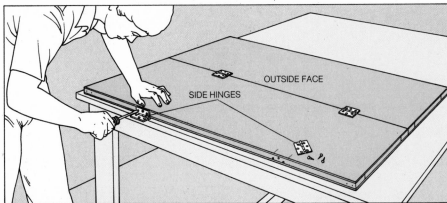

**3** **Mounting the hinges.** Align the panels, outside faces up, ⅛ inch apart. Connect them with two butt hinges equally spaced about 6 inches from top and bottom. Set two more hinges along the side edge of one shutter, aligning them with the hinges in the center and running the hinge pins along the lower edge of the panel, as shown. Screw the hinges to the panel.

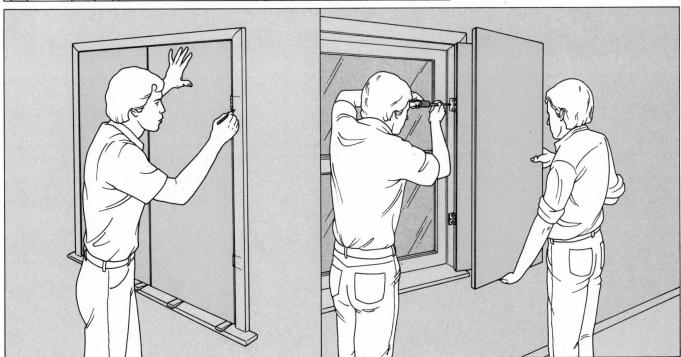

**4** **Hanging the shutter.** Set the closed shutter into the window frame, using wooden shims to position it with a ³⁄₁₆-inch clearance on all sides. Mark the hinge placements on the window casing *(above, left)*. Remove the shutter and transfer the marks to the window jamb. Next, while a helper holds the opened shutter in place, mark the position of the screw holes on the jamb. Remove the shutter, and drill starter holes for the screws. Replace the shutter, screwing the hinges to the jamb *(above, right)*. Use screws long enough to reach a stud—¾ inch for drywall, 1¼ inches for plaster. If necessary, cut a pair of 1-by-1 wood strips the width of the window, and screw them to the top and bottom of the frame to act as stops for the shutter. Attach neoprene weather stripping to the stops and to both of the vertical outer edges of the panels. Screw a knob to the shutter on the unhinged side, and add vertical sliding-bolt latches, if necessary, to hold the shutter closed.

# A Potpourri of
# Window Products

**Multipurpose shades.** Three plastic shades mounted in an insulated triple track control both the gain and the loss of solar heat. On a winter night, all three shades are drawn to form tightly sealed layers of insulation. On a sunny winter day, the clear center shade and the inside heat-absorbing shade are pulled down to trap solar heat. At about 85° F., a thermostat in the lower frame opens a damper, permitting cooler air from the room to pass through vents into the air space between the two shades. As it is warmed by the sun, the air rises and wafts into the room through vents at the top *(inset)*. On a summer day, the reflecting shade blocks hot sunlight from entering the room; on warm nights, all three shades can be raised and the windows can be opened for ventilation.

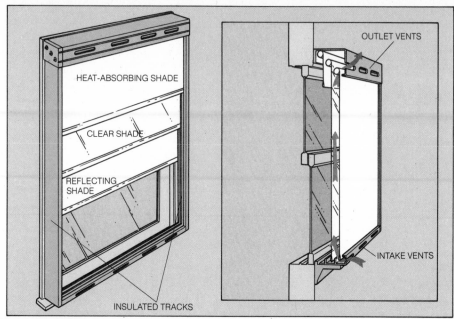

HEAT-ABSORBING SHADE

CLEAR SHADE

REFLECTING SHADE

INSULATED TRACKS

OUTLET VENTS

INTAKE VENTS

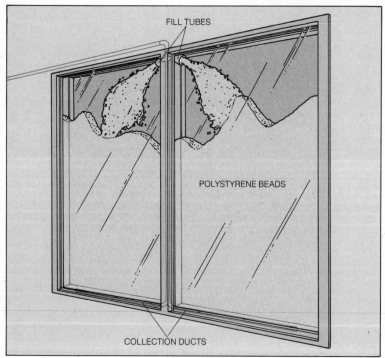

FILL TUBES

POLYSTYRENE BEADS

COLLECTION DUCTS

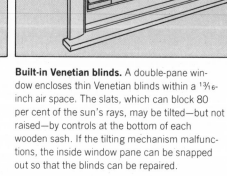

TILT CONTROLS

**A window full of beads.** An ordinary-looking picture window during the day, a double-glazed bead wall is transformed at night: The 2½-inch air space between its two panes of ³⁄₁₆-inch tempered glass fills with millions of tiny styrene beads that increase the window's R value from 1.5 to 8. The beads are blown in through fill tubes at the top of the window by concealed fans activated by either a toggle switch or an automatic heat sensor. In the morning, the sun starts the fans again, but in reverse, creating powerful suction to draw the beads out through ducts in the bottom of the frame and up into a holding tank, leaving the window clear again.

**Built-in Venetian blinds.** A double-pane window encloses thin Venetian blinds within a ¹³⁄₁₆-inch air space. The slats, which can block 80 per cent of the sun's rays, may be tilted—but not raised—by controls at the bottom of each wooden sash. If the tilting mechanism malfunctions, the inside window pane can be snapped out so that the blinds can be repaired.

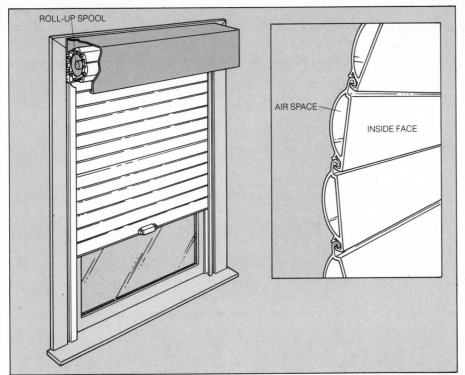

**A shade made of hollow slats.** This indoor blind is made of hollow, interlocking slats of rigid PVC plastic *(inset)* that roll up and down inside an insulated track. The air space inside each slat forms a thermal barrier. The shade is raised and lowered manually on a spring-tensioned spool, as shown here, or mechanically by a small electric motor available as optional equipment from the manufacturer.

ROLL-UP SPOOL

AIR SPACE

INSIDE FACE

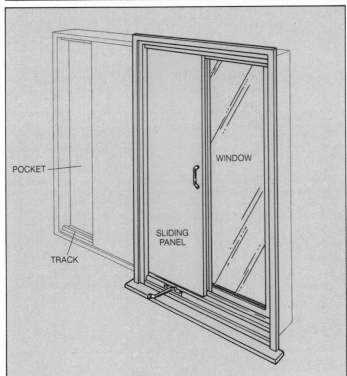

POCKET

WINDOW

SLIDING PANEL

TRACK

**A pocket window.** Set on tracks in the frame of a casement window, a 1½-inch wood-faced panel of rigid foam insulation slides in and out of a 4-inch pocket built into the adjacent wall. The window, panel and pocket are sold as a single prefabricated unit to be installed in a new wall; providing a pocket in an existing wall is an operation that requires extensive alterations.

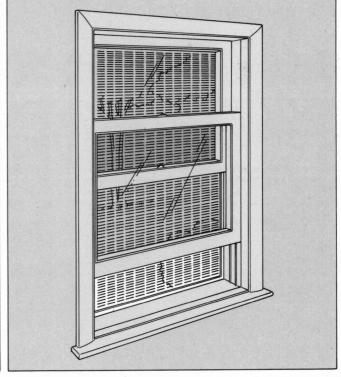

**Solar screening.** Rows of tiny louvers in these aluminum screens block out 80 per cent of the sun's rays, even though the surface of the screen is transparent and 70 per cent open. The solar screening is available in sheets that can be cut to any window size and installed in frames designed for ordinary insect mesh.

# Superinsulation: Thicker Walls to Hold In Heat

Greatly increasing the thickness of house insulation—a technique called superinsulating—is one of the best ways to slash heating bills. In some newly built superinsulated houses—with wall insulation values of R-30 and higher and roof insulation values of R-50 and higher—the savings can be as much as 70 per cent compared with the cost of heating a similar house with conventional insulation. Superinsulating an older house can yield savings approaching these—by stopping drafts and reinforcing inadequate insulation while improving comfort.

Many features found in new superinsulated houses—thermal blocks between structural parts to interrupt the flow of heat, for example, and special roof sheathing—are impractical for existing homes. But other superinsulating techniques are adaptable to existing houses. The primary technique is to bolster existing insulation with thick extra layers. It is simple, for instance, to lay extra fiberglass batting in an attic, adding an R value of about 3.33 per inch of thickness. It is also possible to superinsulate exterior walls.

Before extra insulation can be added, the exterior walls must be thickened to accommodate it. And thickening the walls involves major construction: A wall frame must be built, then covered over with an attractive surfacing material after the new insulation is installed. In almost every case this frame, which can be built of 2-by-4s or 2-by-3s, is best added to the inside of an existing wall. It can rest on the existing floor and can be covered with inexpensive wallboard. Thickening walls on the outside requires a heavy ledger board bolted to the studs to support the new frame, plus expensive exterior siding. However, if your house is due for new siding anyway, superinsulating from the outside may prove feasible.

The major drawback to superinsulating a wall inside is that living space must be sacrificed. A new wall cavity providing room for 6 inches of new insulation will use up 5 square feet of space for every 10 feet of wall thickened; fully superinsulating a two-story, 20-by-40-foot house means the loss of 120 square feet.

Of course, space can be saved by superinsulating only some of the walls. The south wall—warmed by the sun for most of the day—needs superinsulation only in the harshest climates. Or limit your superinsulation to the north wall, or to the north wall and the west or east wall, whichever faces the direction of prevailing winds in your locale.

The major physical obstacles to superinsulating walls are windows, doors, electrical boxes and heating outlets. For windows and doors, use 2-by-2 furring strips to extend the jambs back into the rooms; the recesses inside the extended jambs can be covered with wallboard. Electrical cables connected to boxes, forced-air ducts and radiator pipes must be brought into the new room perimeter.

The vapor barrier of a superinsulated wall also requires special consideration: It must be better than its counterpart in an ordinary wall. Because less air seeps through a superinsulated wall, less moisture is carried out of the wall cavity. Instead, any moisture that gets inside the wall can condense, matting and ruining the new insulation. The best defense is a continuous sheet of 6-mil polyethylene sheeting, spread over the new insulation. The sheeting will bolster the performance of the standard foil vapor barrier that comes with the new insulation.

If there is an existing vapor barrier, it must be perforated so that moisture can pass through it. Otherwise, even a tiny amount of moisture that penetrated the wall would be permanently trapped between two virtually impervious barriers.

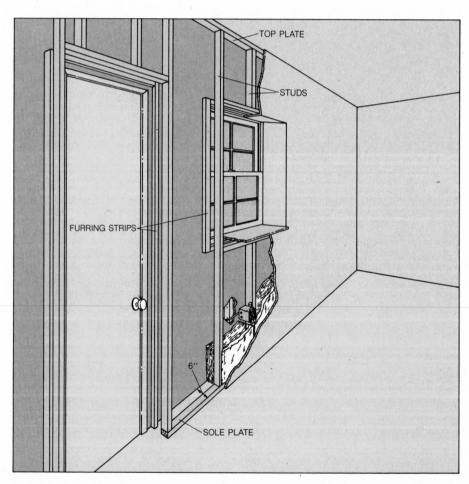

**A superinsulated wall.** An inner frame, consisting of studs, a sole plate and a top plate, thickens an exterior wall from the inside to provide space for 6 extra inches of fiberglass insulation. Window and door recesses in the existing wall are extended inward by the frame and by furring strips, used for securing a new vapor barrier, wallboard and trim. An electrical outlet is relocated onto a stud of the new frame, its front flush with the surface of the new wallboard.

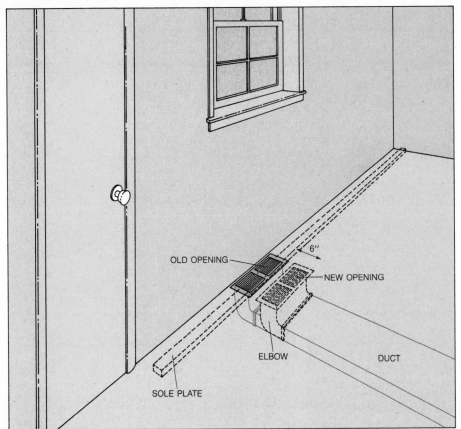

## Moving Existing Utilities

**Detouring a heating duct.** To make room for the sole plate of the new wall frame, move any floor register that is within 6 inches of the existing wall. First lift off the register grille. Working from below, cut away a section of the ceiling under the register if necessary, and pull the elbow free from the end of the duct. Cut the duct short and replace the elbow to reroute the air flow to a new opening cut in the floor. Install the grille over the new opening.

OLD OPENING

6"

NEW OPENING

ELBOW

DUCT

SOLE PLATE

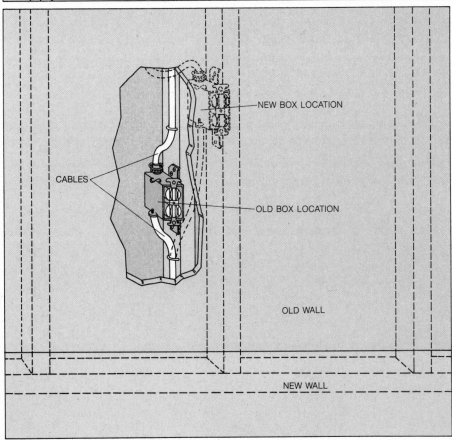

**Pulling an electrical box forward.** If there is enough slack in cables connected to an electrical box, you can avoid having to rewire the wall by simply moving the box from the existing wall to the new inner wall. Turn off the electricity to the box at the house service panel. Hammer a hole through the existing wall surface, and pry out any staples holding the cables near the box. Pull the cables and box loose from the existing wall frame. After securing the new wall frame, remount the box, positioning it so that its front edge will be flush with the new wallboard.

NEW BOX LOCATION

CABLES

OLD BOX LOCATION

OLD WALL

NEW WALL

# Framing the Inner Wall

**1 Adding nailing blocks.** If ceiling joists run parallel to the exterior wall and the new wall frame cannot be located directly under a joist, install nailing blocks between joists to secure the top plate of the new frame. First use a pry bar to lever the baseboards and ceiling moldings off the exterior wall and adjacent walls. If the ceiling is covered with wallboard, use a utility knife to cut it away between the exterior wall and the center of the closest joist. If the ceiling is plaster, cut it away with a mason's chisel and a hammer. Cut nailing blocks from lumber the same size as the joists. Nail the blocks at 24-inch intervals between the joists, with the bottom edges flush.

If the joists run perpendicular to the exterior wall, nailing blocks are not necessary. The top plate of the new wall can be nailed at the points where it crosses the joists.

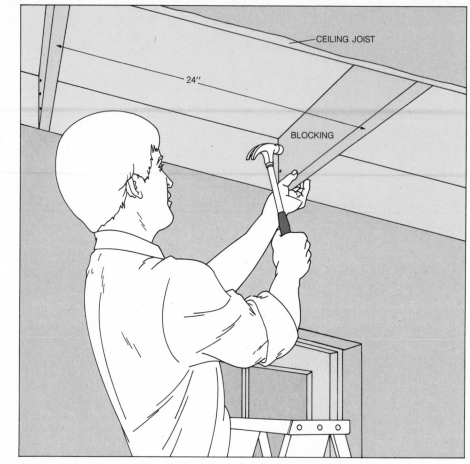

**2 Extending door and window jambs.** Nail 2-by-2 furring strips around the door and window openings to provide a nailing surface for wallboard that will be installed later. To install strips around an untrimmed opening, such as the window shown here, set the furring strips exactly ½ inch from the sides and top of the opening to accommodate the wallboard; at the bottom of a window, nail the furring strip flush with the window sill. For openings trimmed with casing, such as the doorway shown here, space the strips 1 inch or more from the outer edge of the trim.

**3** **Assembling the wall frame.** Cut two boards to serve as the new top and sole plates, the same length as the existing wall. On the face of one of the boards, make marks at 16-inch intervals to indicate stud positions. With a helper, hold the marked board against the wall and mark it for additional studs at each of the vertical furring strips framing the doors and windows. Transfer the marks to the other board. Make two cuts in the sole plate to remove the section corresponding to the doorway, and nail the remaining pieces to the floor.

Cut full-length studs to the height of the room minus 3 inches to allow for the combined thickness of the top and sole plates. Set the marked top plate on edge, and nail a full-length stud to it at each mark, skipping the marks above a door or window. To frame around a window (*inset*), cut two boards the width of the opening. Nail the two horizontally between the studs bounding the opening at the same height as the horizontal furring strips on the existing wall. Nail short studs, called cripple studs, above the header and below the sill at the marked 16-inch intervals.

To frame around a door, nail a horizontal header at the height of the furring strip above the door. Nail cripple studs above the header at the 16-inch marks on the top plate.

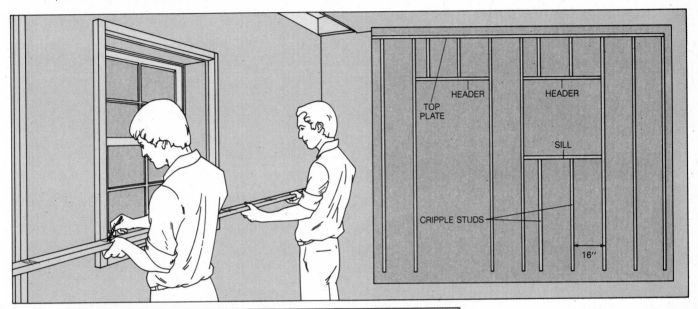

TOP PLATE

HEADER

HEADER

SILL

CRIPPLE STUDS

16"

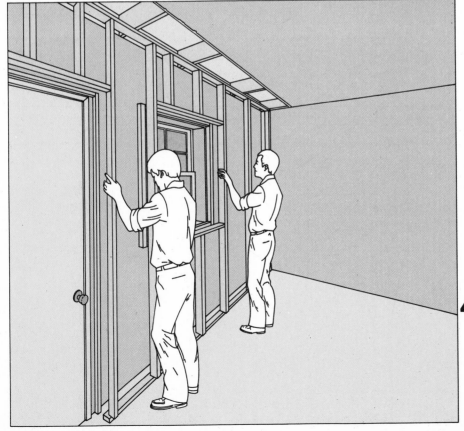

**4** **Erecting the wall frame.** With a helper, lift the assembled top plate and studs onto the sole plate. Toenail the bottom ends of the studs to the sole plate. Plumb the frame, and nail the top plate to the ceiling joists or nailing blocks.

Fasten the end studs of the new wall frame to the adjacent walls with toggle bolts if the adjacent walls are hollow where the frame meets them; nail it in place if the frame falls at stud locations.

# Installing the New Insulation

**1** **Perforating the old vapor barrier.** Using a power drill fitted with a ¼-inch twist bit, bore air holes through the old wall surface at 1-foot intervals vertically and horizontally. Push the drill bit at least 2 inches into the wall to ensure that it pierces the vapor barrier of the existing insulation.

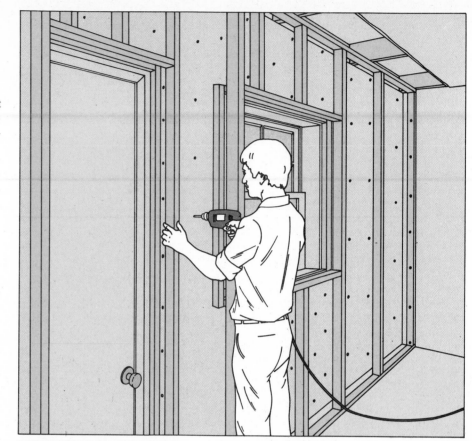

**2** **Stapling on new insulation.** Wearing gloves, a long-sleeved shirt and a dust mask, fit fiberglass insulation between the studs of the new wall frame, keeping the foil vapor barrier of the material toward you. Staple the flanges of the foil facing to the studs every 6 inches with a staple gun. If necessary, cut the insulation to fit with a long pair of scissors or with a butcher knife; odd-sized pieces and scraps can be butted together to fill all of the spaces around doors and windows. Do not pack any of the insulation tight; compressing the fibers will greatly decrease the effectiveness of the material.

**3** **Adding the second vapor barrier.** Fasten a single piece of 6-mil polyethylene over the new insulation, stapling every 6 inches along the studs. At every window and door, slash the plastic from corner to corner in an X pattern *(below, left)*; fold the four triangles into the opening, and staple the plastic to the furring *(below, right)*.

Cover the stud wall and the recesses of the window and door openings with wallboard.

## Finishing the Wall

**Trimming the window.** At the bottom of the window recess, use a piece of stair-tread lumber to create a finish sill. Cut the tread as long as the window recess is wide and ⅜ inch wider than the depth of the recess from the window sash to the new face of the wall. Place the back edge of the tread 1/16 inch from the window sash, and nail through the top of the tread into the framing and furring strip below. Cover the seams between wallboard panels here and elsewhere in the room with joint compound and tape.

# A Storage System for Solar Heat

An indoor thermal storage system can reduce winter heating bills by trapping sunlight as it enters your house and storing the heat from the sunlight until it is needed. The principle behind thermal storage is simple: A solid substance or a container of liquid placed beside a window in direct sunlight absorbs heat from the sun's rays. When the air surrounding the heat-storing mass becomes cooler than the surface of the substance or of its container, the material begins to release the stored heat.

Putting the principle of thermal storage into practice is also simple, provided that several conditions are met. Thermal storage systems work best in rooms that have their windows facing south. The more windows the better; as a general rule, the glazed surface area should equal at least 10 per cent of the floor space in the room. Such rooms are prone to overheating as the sharply angled rays of winter sunlight pour through the windows during the day; it is this tendency to overheat that makes the room a good candidate for a thermal storage system. The added mass near the windows absorbs the extra heat before the room has a chance to become too hot; the stored heat is not released until it is needed—after the sun goes down.

You can perform an easy test to determine whether added thermal mass in a room or section of your house will store and radiate enough BTUs of heat to help cut heating bills. On a sunny winter day, pull back drapes and blinds to uncover all the windows in the room, then place a thermometer in the center of the area. If, between 10 a.m. and 2 p.m., the temperature in the room reaches 90° or higher, thermal storage will work efficiently and cut heating costs.

Masonry, concrete, water and a liquefied chemical compound called phase-change material (PCM) are all used to provide the mass required for thermal storage. Some of these materials are more practical for new construction than for a solar retrofit.

Masonry and concrete, for example, are generally installed as part of the floor area adjacent to windows or as columns partially blocking windows; these materials are most efficiently utilized when they are incorporated in the design of a room. Certain water and PCM containers are made to be built into the structure of a wall; these are also easier to install during rather than after construction.

The water and PCM containers illustrated on these pages are designed for use as part of a solar retrofit. The cylindrical tanks (opposite, below) are made of fiberglass and are available in 12- or 18-inch diameters, in heights of up to 10 feet. They hold up to 132 gallons of water. You can add dye to the water to

## Determining Storage Needs

**Calculating quantities of thermal mass.** To use the chart at right, first total the square feet of floor space in the room or area to be heated; then total the square feet of south-facing glass in the same room or area. In the left-hand column of the chart find the description that best defines the type of insulation in your house. Read over to the next column and choose the temperature closest to the average January-February temperature in your region. Find the conversion factor directly to the right of the temperature and multiply the floor-space figure by this number.

If the result of your multiplication is greater than the south-facing-glass figure, thermal storage will not work efficiently in the planned area. If the result of the multiplication is less than the south-facing-glass figure, multiply the difference by the thermal-mass base quantity for the storage material you plan to use. The final figure will tell you how much thermal mass to add for optimum heat storage in your house.

| House insulation | Average January–February temperature | Conversion factor |
|---|---|---|
| **Standard:** 3½" fiberglass batts or equivalent in 4" walls; 6" fiberglass batts or equivalent in ceilings; double-pane glass; weather-stripped windows and doors. | 20°F. | .115 |
| | 30°F. | .105 |
| | 40°F. | .90 |
| **Heavy:** 6" fiberglass batts or equivalent in 6" walls; 9" fiberglass batts or equivalent in ceilings; 3½" batts in floors or 1" styrene or urethane insulation in basement; triple-pane glass or double-pane glass with night insulation; weather-stripped windows and doors. | 20°F. | .92 |
| | 30°F. | .84 |
| | 40°F. | .72 |
| **Super:** 6" fiberglass batts or equivalent in 6" walls; 1" styrene or urethane insulative wall sheathing (sub-siding); 12" fiberglass batts or equivalent in ceilings; 6" batts in floors or 2" styrene or urethane perimeter insulation in basement; triple-pane glass or double-pane glass with night insulation; weather-stripped windows and doors. | 20°F. | .83 |
| | 30°F. | .75 |
| | 40°F. | .64 |

**Thermal-mass base quantities**
Water: 7 gal. (56 lb.)
Phase-change material (PCM): .8 gal. (10 lb.)
Brick: 5 sq. ft. (217 lb., 38 bricks)
Concrete slab: 4 sq. ft., 4 in. thick (200 lb.)
    3 sq. ft., 6 in. thick (200 lb.)

darken it and thus increase its absorptive capacity, and algicide to retard the growth of organic matter.

The tanks are deceptively light when empty—the fiberglass in each cylinder weighs only about 10 pounds; however, an 8-foot-tall, 12-inch-diameter tank, for example, weighs 380 pounds when filled with water. Unless you can set the tanks on a floor that covers a concrete slab, local building codes, which you should check before beginning the installation, will generally require that you set each tank directly over a floor joist that is reinforced by the method shown on page 34, Steps 1 and 2.

The rest of the installation is simple: 1-inch-thick Manila or cotton rope looped through 1¼-inch eye screws secures the tanks at the top (do not use nylon rope because it stretches). The kickplate tray underneath the tanks is optional; if you want one to provide a flat surface and to protect the bottoms of the cylinders, you can fabricate it with 30-gauge, galvanized sheet steel.

The PCM pod strips on page 35, although considerably more expensive than water tanks, take up one quarter the amount of space and store more than five times the BTUs of heat per pound of material. (The enhanced heat-storing capability of the pod strips is due to the phase-change material—a saline compound—sealed within the fiberglass pods of each strip; this chemical changes from a solid to a liquid at 81°.) Because they weigh no more than 29 pounds apiece and slip readily in and out of their aluminum support channels, you can take them out in the summer, when heat storage is undesirable.

Both the solar water tanks and the PCM pod strips with the necessary hardware for installation are obtainable from solar-equipment dealers; if you have difficulty finding them in your area, check with the U.S. Department of Energy for a list of manufacturers.

Before you install any type of thermal storage, use a simple mathematical formula in conjunction with the chart shown below, opposite, to determine how much mass you must add to the area that will be heated. Your calculations will be based on several factors. First, it is essential for you to know how well your house is insulated—use the descriptions in the chart for guidance. If your house does not meet the standards listed as average in the chart, you should add the required insulation before installing thermal storage materials.

The amount of thermal mass you add also depends on the ratio between the number of square feet of glazed surface in the room and the number of square feet of floor space to be heated. You must know the average January-February temperature in your region; check this by contacting the National Climatic Center in Asheville, North Carolina.

Finally, the amount of mass depends on the type of storage material you are planning to add. The table underneath the chart translates your calculations into an appropriate volume of water, phase-change material, brick or solid concrete.

**Cylindrical tanks and sealed pod strips.** Anchored 4 inches from the stationary section of a sliding glass door, three water-filled translucent fiberglass cylinders (*above, left*) trap and store the solar radiation transmitted through the glass of the door. At night, when the room temperature drops below the temperature of the water in the tanks, the thermal energy stored

in the water radiates through the fiberglass walls and heats the air in the room.

The considerable weight of the cylinders is supported by reinforced joists beneath the floor (*page 34, Steps 1 and 2*); the tank bottoms rest in a sheet-metal tray, which provides a smooth, level surface as well as a 2-inch-high protective

kickplate. The top of each cylinder is secured to the wall with rope looped through eye screws. Sealed fiberglass pod strips containing phase-change material (*page 33*) cover large windows along a south-facing wall (*above, right*). Resting in aluminum slip channels supported by wall brackets at each end, the pod strips can be removed easily during summer months.

# Fiberglass Water Tanks for Thermal Storage

**1** **Preparing joists for reinforcements.** For each joist that needs reinforcement (*page 33*), cut enough 18-inch spacer blocks from matching lumber to run, spaced at 2½-foot intervals, along both faces of each joist. Nail each block to a joist face with three 12-penny nails, staggered in a zigzag pattern as shown.

Cut two reinforcement joists for each joist; use matching lumber and make each reinforcement the same length as the existing joist. Miter the ends of each new joist at opposing 60° angles; this makes them easier to slip into place.

**2** **Installing the reinforcement joists.** With its longer edge facing down, tilt a reinforcement joist into place beside an existing joist; rest the ends of the new joist on the sill plate and push the face of the joist against the spacer blocks. Nail the reinforcement to each spacer block with three 16-penny nails, reversing the zigzag pattern used for the blocks in Step 1, above. Nail a second reinforcement to the blocks on the opposite face of the joist; repeat the same procedure for every joist that needs reinforcement.

**3** **Anchoring the water tanks.** For each water tank, drive two 1¼-inch eye screws, spaced at a distance equal to the diameter of the cylinder, into the header beam above the window. Leave at least 2 inches of space between screws for adjacent tanks. Then set one cylinder in place, 4 inches from the window with its bottom end in a kickplate tray (*page 33*). While a helper checks with a level to be sure that the tank remains plumb, loop a 1-inch-thick rope snugly around the tank and through the pair of eye screws, securing the ends of the rope with double square knots. Anchor the remaining tanks in the same manner.

Use a garden hose to fill the tanks with water to within 3 inches of their tops; add any desired dyes or algicides and then push the cap onto the top of the cylinder.

# Heat-storing Pods
# Installed in Strips

**1** **Installing support brackets.** Set the predrilled flange of a channel-support bracket against the wall beside the window, the lower edge resting on the window stool and the corner of the bracket flush with the casing. Make marks on the wall at the screw-hole locations and along the top edge of the bracket. Measure 15⅝ inches up from the top-edge mark and set the top edge of a second bracket at this point; with the corner of the bracket against the window casing, make the same marks as before.

Continue up the wall making similar pairs of bracket marks—one lower and one upper—for each pod strip you will install. Then mark the other side of the window in the same way. At the marks, screw the brackets to the jack studs on either side of the window, using 2-inch wood screws. For drywall with no framing member underneath, use Molly bolts; for masonry, use lead anchors and screws.

Measure the horizontal distance between the protruding flanges of two brackets; use a hacksaw to cut strips of aluminum channel 1½ inches longer than this measurement; cut one strip for each horizontal pair of brackets.

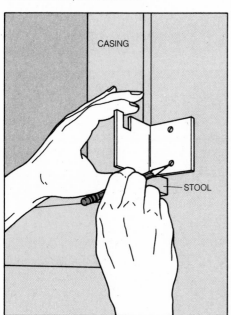

CASING

STOOL

**2** **Setting the pod strips in place.** Rest each aluminum channel strip in the bracket slots at opposite sides of the window (*below*). To install each pod strip, slip the pod's upper edge into a channel groove, pushing up until its lower edge will clear the bottom lip of the channel below, then drop the lower edge of the pod into the bottom lip (*inset*).

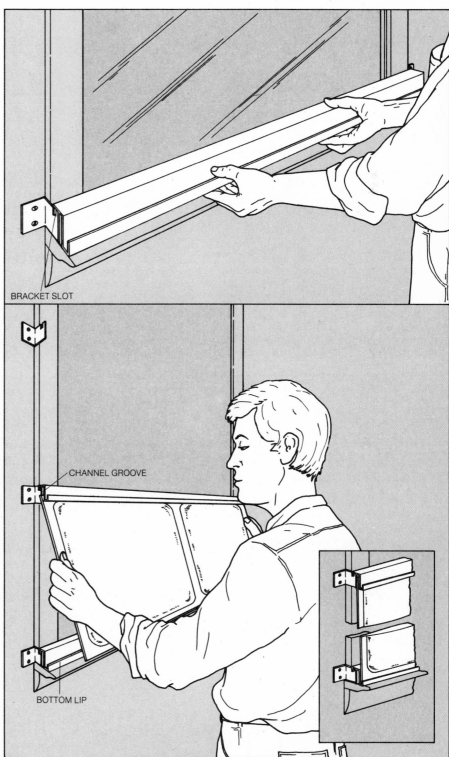

BRACKET SLOT

CHANNEL GROOVE

BOTTOM LIP

# Portals That Lock Out Cold Air

A classic way for keeping cold air from leaking into a house, the double-doored entryway is back in fashion as a device for conserving energy. Known in past incarnations as a vestibule or foyer, it is now called by energy experts an air lock—and when tightly weather-stripped will reduce drafts, a major cause of heat loss, by as much as 15 per cent.

Though credited with substantial savings in the cost of heat, an air lock makes most economic sense when it does double duty as a mud room for a household of children, a pantry for a crowded kitchen, or a convenient coat closet. Purely as an energy-saving device, it may take more than 10 years to pay back the cost of construction; but the bonus of extra space may tip the scales in its favor.

There are many different ways to build an air lock, depending on the floor plan and architecture of your house. The simplest method, in houses with entrance halls, is to wall off a section of the hall *(pages 26-31)* and install a second door *(below)*. It is almost as easy to enclose a small entrance stoop, which typically consists of a broad concrete slab shielded by an identically sized roof. Permanent walls of wood studs are erected between the roof and the step, and a new exterior door is set into a rough frame in one of the walls *(opposite, top)*. The outside of the stud wall is then covered with siding, the inside with wallboard.

A removable air lock can be built on the same type of entrance stoop, using detachable panels of aluminum and glass. These panels are storm doors, ordered without handles or hinges. Special connecting hardware—metal channels and angles—can be ordered from suppliers to convert the panels into wall sections. In spring, the glass in the panels can be replaced with screens, or the whole enclosure can be dismantled.

Whatever form of air lock you choose, plan to cover a floor space at least 4 feet square, the minimum area needed for comfortable maneuvering. Local building codes may dictate an even larger floor area. Within this space, the positioning and swinging of the two doors are critical design considerations: The doors should not open into each other, nor should they need to be open simultaneously for a person to pass through the air lock.

To stop drafts, both doors must be weather-stripped. For wooden doors, energy consultants recommend attaching wooden strips that come with a hollow vinyl bead set into the edge. The strips are cut to fit the top and sides of the jamb, then nailed in place. A metal door sweep with a vinyl strip is used to seal the bottom edge of a wooden door. Metal storm doors come with their own weather stripping attached.

Providing light in the confined space of an air lock can be a special problem. Sometimes you can avoid the need to add electric lighting by using glass doors or installing a window in a wall adjacent to outdoor lighting. More likely, however, you will have to wire a lighting fixture into the ceiling of the air lock and a light switch into a nearby wall, tapping an existing outlet for electricity.

## Three Approaches to an Air Lock

**An inner door in an entrance hall.** Sealed off by a new interior door, an existing entrance hall forms an airtight passageway. Jack studs, cripple studs and a header enclose the rough frame for a new prehung door; studs, header and frame are all constructed of 2-by-4s. The door, shimmed into the frame after the new wall is covered with wallboard, is a split-jamb unit, so called because the two halves of the jamb fit together with a tongue-and-groove joint *(inset)*. The casing comes attached to both jamb halves.

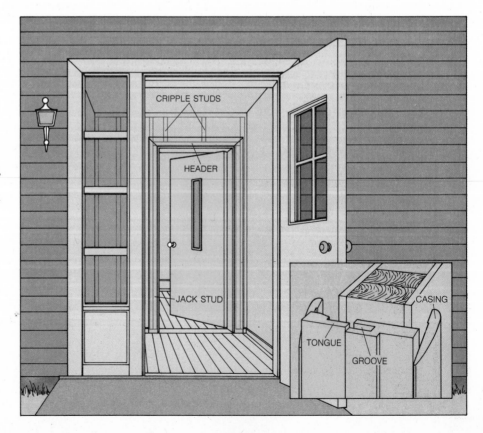

CRIPPLE STUDS

HEADER

JACK STUD

CASING

TONGUE

GROOVE

**A permanent outdoor air lock.** Three short stud walls, one fitted with a prehung door, protect an existing entrance door from drafts and provide a convenient storage place for muddy boots and outdoor clothing. The 2-by-4 wall frames—made of vertical studs and horizontal top plates and sole plates—are nailed into the floor of the concrete stoop, to the joists of the overhanging roof and, at the house wall, to the existing studs. Matching pieces of siding, sheathing and wallboard cover the new walls. A glass-paneled door and a window admit natural light into the enclosure during the day, so that electricity is required only at night.

**A removable porch enclosure.** Aluminum storm-door panels, held by a frame of interlocking aluminum channels, turn a small entry porch into a weatherproof enclosure that can be disassembled and stored. The side panels are set into F-shaped channels *(inset)*, which are screwed into the house wall, the concrete stoop, the porch roof and, at the front, square aluminum posts. The door is installed between a second set of aluminum posts. To prevent drafts, all the panel edges are caulked.

# 2

# Harnessing the Heat of the Sun

**Solar building materials.** Sturdy aluminum bars, which can be cut to size with a circular saw, are assembled like the pieces of a picture frame to hold a ribbed acrylic panel. The transparent, double-skinned panel transmits light as well as glass does, but the acrylic is far tougher, making it an excellent glazing material for solar collectors. A flexible silicone-coated gasket fitted into the frame allows the panel to expand and contract without leaking when the temperature changes.

The slim, black solar collector panels fast becoming a familiar sight in communities across the country are not a novelty born of present-day fuel shortages. Americans began economizing with solar collecting devices before the turn of the century. A simple solar water heater similar to the batch heater on page 89 was patented in 1891. It featured four galvanized-iron water tanks, painted black and mounted inside an insulated pine box with a glass cover. By 1900, the Climax, as it was called, crowned thousands of American roofs. But its popularity did not last: The advent of cheap natural gas made solar water heaters passé by the 1930s.

The designers of the Climax were not the first to learn that a space enclosed with transparent glass would trap the heat of the sun. Centuries earlier, the comfort-seeking Romans who basked in steamy, glass-walled bathhouses were utilizing the main principle of solar collection: the greenhouse effect.

Sunlight is made up of very short wavelengths of electromagnetic energy, a characteristic that permits it to pass easily through transparent materials such as glass. If the light then strikes a dense, opaque substance, it is absorbed and turned to heat. But the heat waves, being longer than light waves, cannot readily pass back through the transparent material. Thus, a transparent enclosure becomes a heat trap. The modern, flat plate solar collector works exactly that way: In effect, it is a small greenhouse.

Solar collectors are at the heart of each of the solar heating devices on the following pages. Active solar systems employ mechanical means such as a fan or pump to move the collected heat to where it is needed. Others, called passive systems, rely on the natural principles of heat transference to do the job.

Heat always attempts to flow from a warm area to a colder one. The three methods by which it moves are radiation, conduction and convection. Radiation is the movement of heat waves across space; anything warmer than its surroundings radiates heat to the objects around it. Conduction is the movement of heat through a substance by molecular vibration. The solid masonry Trombe wall shown on page 41, warmed on the outside by the sun, conducts heat through itself to the cooler house interior and radiates it into the room.

Convection is the natural circulation of heated air. When the air is warmed it expands, grows lighter and rises. As it loses heat to its surroundings, it grows denser and sinks. The air heated in a wall-mounted thermosiphoning air panel *(page 49)* travels by convection. It rises and escapes through an upper vent into the house, where it gives up its heat, falls toward the floor, and exits the house through a lower vent, completing a circular current of air.

# The Trombe Wall—A Solar Heater Made of Masonry

The Trombe wall—named after its inventor, French architect Felix Trombe—converts a solid masonry wall into a passive solar collector that can provide about half of the required heat for the living space it adjoins. Mounted on a wood frame 3 inches out from the house wall, transparent panels of glass, fiberglass or plastic trap solar radiation in a narrow hot-air sandwich. The sun's rays heat the masonry; the sun-baked brick, concrete or stone then warms the room inside. When vents are cut in the house wall, convective airflow augments the radiant heat by circulating drafts of heated air through the vents directly into the house.

The principle is simple, but not every wall can become a Trombe wall: Certain conditions must be met before you consider this type of solar retrofit. The wall must be unshaded during the winter and shaded—either by trees or by an awning—in summer, and it must face within 20° of true south. It must be solid masonry—brick, concrete, stone or adobe—at least 8 inches thick, with no insulation behind it and no internal air pockets. Such pockets, often left during construction to provide thermal insulation or to prevent moisture build-up within the wall, prevent the masonry from storing and radiating heat efficiently. Check your building blueprints for evidence of air pockets or, if the blueprints are not available, tap the wall over its entire surface for telltale sounds of hollowness.

Finally, check the condition of the wall's outer surface. The masonry must withstand temperatures that can range from 140 to 180° F., so be sure to replace broken bricks or stones and repair crumbling mortar joints.

Before building a Trombe wall, you should also check with the local building inspector and with any zoning commission or architectural review board that may exist in your community. The masonry surface of a Trombe wall is often painted black for maximum absorption and retention of heat. The wall works well enough without the paint, however—which is fortunate because in some communities the black façade, coupled with its cover of transparent panels, may be considered unsightly and, for this reason, undesirable or even prohibited.

Once these preliminary requirements have been met, you are ready to plan the design of the wall. Trombe wall panels come in a standard 4-by-8-foot size and are available from solar-equipment dealers. Plastic and fiberglass panels are generally more popular than glass because they are light and easy to work with. They come with rubber gaskets at the top and bottom; be sure to remove the bottom gaskets before you begin the installation. Glass panels, though heavy and fragile, are also a possibility; all types of panels have double walls for maximum heat retention.

You can buy all of the necessary framing materials for the panels from the same source; they come in plastic, aluminum or wood, along with the appropriate gaskets, depending on the type of panel. To save money, you can buy pressure-treated lumber and cut the framing pieces yourself, as shown opposite, and substitute a black rubber called neoprene for the preformed gaskets; it comes in strips and rolls from hardware stores.

For optimum efficiency, the area of the wall should equal approximately one third the floor area of the room the wall is expected to heat. But in planning the wall, take the size of the panels into consideration. It is usually easier to adjust the dimensions of the wall than to cut numerous glass or plastic sheets to odd sizes. Ideally, the only time you should cut panels is when you plan to frame around a window, either to provide a fire exit or to avoid interfering with a view, as in the case of the right-hand window shown opposite.

A Trombe wall can be designed to function with or without vents, depending on the time of day when heat is most desired. An unvented wall, heating by radiant heat alone, begins its heating cycle late in the day and continues to provide heat into the night. A vented wall, heating by convection as well as radiation, begins its heating cycle earlier in the day—as soon as the sun warms the air trapped under the glazing.

The openings for a vented Trombe wall should always be installed in pairs, in vertical alignment, one just above floor level, the other below ceiling level. The size and number of vents is determined by the size of the Trombe wall they will service. The combined area of the vents should be approximately equal to 1.5 per cent of the total Trombe wall area.

To calculate the size of the vents, first calculate 1.5 per cent of the total area; then divide this figure into an even number of vents. Each vent should be about three times as wide as it is high; no vent should exceed 20 inches in width. If the Trombe wall will cover a double-hung window, you can use the upper portion of the window for a top vent, as shown opposite, and adjust the size of the other vents accordingly.

The installation of a Trombe wall is a fairly simple job, but one that requires painstaking measurements. An error of even as much as ¼ inch in the placement of a framing piece can make it impossible to complete the installation, since glass and plastic panels are quite difficult to trim accurately.

At the same time, the framing must allow for slight expansion and contraction of the panels caused by changes in temperature—requirements that will be specified by the panel manufacturer. The measurements given in the Trombe wall instructions on the following pages are for acrylic panels set into a wood frame. If you install a different framing system or another type of panel (pages 48-57 and 64-77), refer to the installation instructions that accompany these materials for the correct measurements.

For this installation, you will need only a few basic tools—a circular saw for cutting the wood framing and plastic panels; a hammer drill, which can be rented from a tool-rental store, if you will be cutting vents in the masonry; and a socket wrench for fastening expansion bolts to the masonry wall.

**Anatomy of a Trombe wall.** This Trombe wall is made up of six 4-by-8-foot panels of double-walled acrylic *(inset, bottom right)* mounted on a rectangular frame of doubled 2-by-4s that allows a 3-inch air space between the panels and the masonry house wall. The 2-by-4 framing members are fastened to the masonry wall with expansion bolts *(inset, bottom left)*. The outer edges of the panels at each end of the wall are sandwiched between ⅛-inch-thick strips of neoprene gasket, butted against 1½-by-⅝-inch separator strips and covered by 1-by-4-inch battens; this

entire assembly—gaskets, separator strips and battens—is screwed to the framing.

Interior panel edges butt against separator strips mounted on 2-by-4 mullions. The mullions are offset from the house wall by spacers made of 1½-inch lengths of PVC pipe, through which run expansion bolts that fasten the mullions to the masonry. The joints between interior panel edges are covered by 1-by-4 battens; screws driven through the centers of the battens into the separator strips anchor the joints firmly.

Here, the left-hand section of the Trombe wall is vented; three vents are cut into the masonry, and the fourth vent is the top of a double-hung window. The right-hand section is unvented and its window is left uncovered, providing a fire exit. The window is surrounded by a frame constructed of doubled 2-by-4s in the same manner as the perimeter framing, and the plastic panels are cut to fit around the window frame. The two sections of the wall—vented and unvented—are sealed off from each other by a mullion made of doubled 2-by-4s.

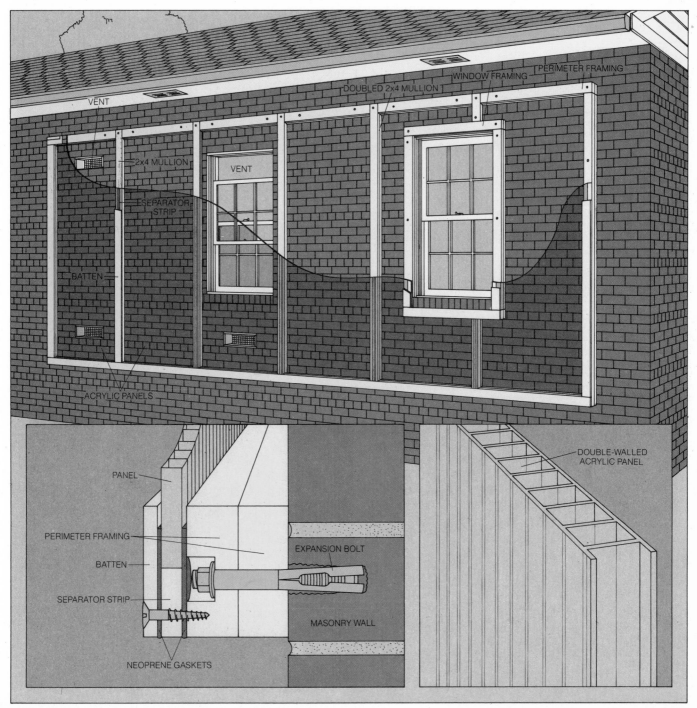

**How a vented Trombe wall works.** The diagram at near right illustrates the daytime pattern of natural circulation that keeps warm air flowing from a vented Trombe wall into the adjoining living area. The air in the space between the glazed panels and the masonry wall is heated by the sun-warmed masonry. The heated air rises, creates a current strong enough to push open a flexible damper flap that covers the upper vent, and flows into the house. At the same time, cool air from the room is drawn through the lower vent, pushing open a similar damper flap, and is heated to continue the cycle. This convective circulation continues until two or three hours after sunset, when the masonry wall becomes too cool to produce an updraft of warm air.

At night, the air in the wall space cools, tending to drop and flow into the room through the lower vent. This change in the direction of airflow, which would draw warm air out of the room through the upper vent, is checked by the flexible damper flaps, which are drawn closed against a screen or a register by the reversed circulation pattern. But the masonry wall still radiates heat it has absorbed during the day (*far right*).

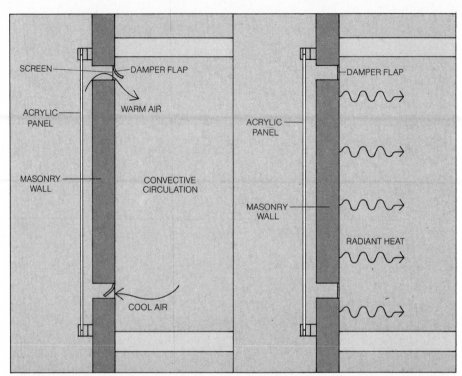

**Planning a Trombe wall frame.** The dimensions of a Trombe wall frame are planned to accommodate the parts it will hold—the glazed panels and the separator strips against which the edges of the panels are butted; these strips may vary in width with the size of glazing used. In addition, the expansion allowance needed for the panels must be taken into account. Usually this allowance is ¼ inch per panel—⅛ inch on each side—but manufacturers may specify more.

To determine the length of the top and bottom frame sections, add together the width of all the panels, the width of the separator strips at all the mullions, and the width of the separator strips along each side of the frame; then multiply the number of glazed panels by the expansion allowance, usually ¼ inch, and add this to the result (*right*). To calculate the overall height of the mullions and side sections of the frame, add the height of one panel to the width of two perimeter separator strips, plus the expansion allowance. For the side sections, subtract the combined width of the top and bottom frame sections—usually 7 inches, since most 2-by-4s are 3½ inches wide—from the overall height.

To determine the position of a mullion on the plan, locate the separator strip for the mullion, using the panel width and expansion allowance as a guide. Then locate the strip's center line, and mark that center line on the top and bottom frame sections. On each side of this mark, measure out half the width of the mullion (*inset*).

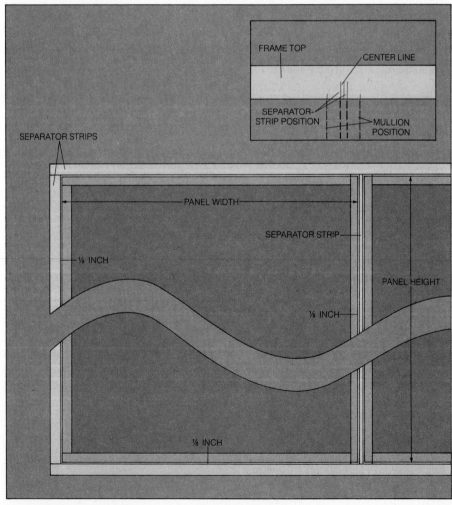

## Constructing the Vents

**1 Breaking through the brick.** Prepare the indoor space for dust and debris: Pull back furniture from the wall, cover the floor and furniture with dropcloths, and shut room doors. Mark the vent locations on the outside wall; usually they should be centered between mullions. Then, wearing protective goggles and gloves, use a 3-pound mallet and a cold chisel to cut into the mortar joints along a vent outline. Try to remove whole bricks within the outline in one piece, but break them if you have to. To split cleanly through bricks that are crossed by the outline, score them along the outline with a brickset (*inset*), then cut along the scored line with a brickset or cold chisel. Repeat for every vent.

After the outer course of bricks has been removed, drill a hole through the center of each vent with a carbide-tipped bit. Working from inside the house, use the holes as reference points to mark the vent outlines, and remove the inside bricks in the same manner.

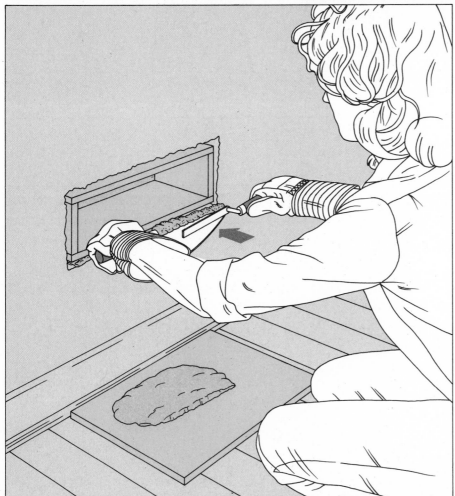

**2 Installing the sleeve.** For each vent, construct a box of ¾-inch wood, the same depth as the vent and dimensioned to fit snugly inside it. Fit the boxes into the vents and, wearing gloves, stuff fiberglass insulation into the gaps around the edges. Cover the fiberglass, on both inside and outside walls, with a ½-inch-wide line of mortar. Block each vent temporarily, while you finish the Trombe wall, with a piece of rigid foam insulation, wrapped in polyethylene to simplify its removal.

# Installing a Trombe Wall Frame

**1 Establishing string guides.** At one of the upper corners for the Trombe wall location, drill a 1-inch-deep hole into a mortar joint and tap in a 4½-inch nail; make sure that the nail is perpendicular to the wall and projects from it approximately 3½ inches. Mark the planned height of the frame on the string of a plumb bob, and suspend the plumb bob from the nail; using this for reference, make a mark for the lower corner of the Trombe wall (*below, left*). Then measure the planned length of the wall along the mortar joint holding the nail, and mark the other upper corner of the Trombe wall. Hold the plumb bob against that corner, and mark the corner below it. Check to make certain that the four corners form a perfect rectangle by measuring the diagonals between them, adjusting the marks as necessary until the diagonals are exactly equal.

Drill pilot holes and drive nails into the three marked corners. Tie a string near the head of one nail and run the string taut around the others, fastening it to the first nail. At 1-foot intervals, measure the distance between the string and the wall, and slide the string along the nails until the minimum distance between string and wall is exactly 3 inches (*below, right*). Paint the area within the string with flat exterior latex paint.

**2 Framing the perimeter.** Assemble doubled 2-by-4s for the perimeter frame, leaving 3-inch gaps in the outer 2-by-4s of the top and bottom sections for the mullion joints. Hold the lower edge of the bottom frame section to the guide string and, at one point where the inner 2-by-4 touches the brick, drill through the frame and into the brick. Apply construction adhesive to the wall side of the inner 2-by-4, and bolt the frame section to the wall with an expansion bolt (*page 41, bottom left*). Add other bolts at 3-foot intervals; remember that mullion ends will also need bolts. Use shims, if necessary, to hold the face of the outer 2-by-4 flush with the string guide.

Bolt the side and top frame sections to the wall. If you are using glass panels, add a brace as on page 53 to the wall and bottom frame piece. Trim the shims, stuff fiberglass into any gaps between the frame and the wall, and seal the perimeter with silicone caulk. To build around a window, frame it with doubled 2-by-4s, working in the same manner as for the larger frame.

GUIDE STRING

**3** **Affixing the mullions.** Drill mounting holes in the mullions as for the perimeter framing. Position a mullion in its gaps in the top and bottom frame sections, and bolt it to the wall. Install the remaining mullions in the same way. If you are constructing vented and nonvented areas within a single Trombe wall, separate them with a doubled 2-by-4 mullion to prevent air from flowing between the two sections. To construct the doubled mullion, add a 2-by-4 the length of a side piece to a 2-by-4 the length of a regular mullion.

**4** **Attaching the spacers.** At 3-foot intervals along each mullion, hold a piece of ¾-inch diameter PVC plumbing pipe against the house wall; mark where the pipe touches the inside face of the mullion. Cut the pipe at this mark. Then drill a bolthole through the center of the mullion and a corresponding hole in the wall. Insert an expansion bolt through the mullion, through the PVC spacer and into the wall. Tighten the bolt.

# Adding Panels and Flashing

**1** **Assembling the separator strips.** Drill ⅛-inch-diameter weep holes through the separator strips that will be fastened to the bottom frame section; usually 1 to 3 weep holes per panel is sufficient, but check the panel manufacturer's instructions for recommended intervals. Sandwich a gasket between each bottom separator strip and the bottom frame and, holding the bottom edge of the strip flush with the bottom edge of the frame, nail the strip to the frame with 1½-inch nails every 6 inches. Continue in this fashion, attaching separator strips and gaskets to the sides and top of the frame, but do not drill weep holes into these strips. In the same manner, attach gaskets and separator strips along the centers of the mullions. As each raised frame of separator strips is completed, check its interior dimensions to be sure a panel will fit into it.

**2** **Securing the panels.** With a helper, hoist an end panel into place within its frame of separator strips. While your helper stabilizes the panel, hold another gasket against the edge of the panel, and top the gasket with a batten. Be sure that the gasket lines up with the outer edge of the batten and that the outer edge of the batten lines up with the outer edge of the frame. Screw through the batten into the separator strip every 2 feet, using 3-inch wood screws. Continue installing panels in this fashion until all the panels and battens are in place.

When panels must be cut, as for fitting around a window, use the tool recommended by the manufacturer; for most plastic panels, this will be a circular saw fitted with a plywood-cutting blade. Apply masking tape to panel areas along which the saw will travel, to prevent the saw base from scratching the panel. Use gaskets to cap the bottom edges of any cutout in order to help keep out moisture.

**3** **Installing the flashing.** At the mortar joint immediately above the top of the Trombe wall, use a circular saw with a masonry blade to rout a ¾-inch-deep groove along the entire length of the wall. Fashion—or have a fabricator fashion—a strip or strips of aluminum flashing wide enough to fit into the groove, project over the Trombe wall and bend over the wall's front edge about 1 inch. Fill the groove with mortar and insert the flashing edge; hold the flashing there until the mortar stiffens.

## Finishing the Vents

**Adding the draft flaps.** Working inside the house, remove the rigid foam insulation from the vents. Cut a piece of the wire mesh called hardware cloth to fit the exterior dimensions of an upper vent, and staple the mesh to all four edges of the vent. Cut a 2-mil polyethylene flap to fit the interior dimensions of the vent—but add ⅜ inch to the length and width of the flap. Bend duct tape over the top ½ inch of the polyethylene for reinforcement, and staple the flap to the top edge of the vent *(left, top)*. Add a flap and screen to each upper vent in the same fashion.

For a lower vent, first staple a polyethylene flap, cut ¾ inch less wide than the draft flaps for the upper vents, to the edges of the vent; then add the mesh *(left, bottom)*. Frame the vents with molding of your choice; here, clamshell molding is used *(inset)*. Fasten the outer portion of the molding to the wall surface with a construction adhesive, and nail the inner portion to the edges of the vent with 1½-inch finishing nails.

As an alternative to this arrangement of screening and draft flaps, construct the vent cover shown on page 57, or slip store-bought air-duct registers into the vents and screw the registers to the wall. Omit the molding—the register flanges cover the sleeve sides. Open and close the registers night and morning to control the direction of the airflow.

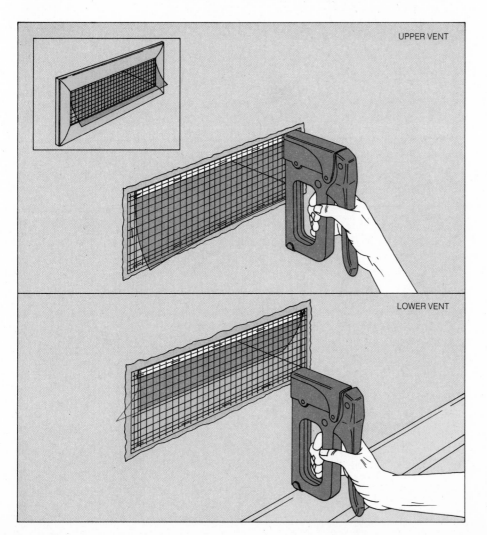

UPPER VENT

LOWER VENT

# Warming a Room with a Glass-faced Box

A thermosiphoning air panel, commonly called a TAP, is a single-room space heater that uses convection to waft warm air into a room. A TAP draws cool air near the floor of the room into the bottom of a glass-faced box. There, the air is warmed by a metal absorber plate that itself is heated by the sun; the air then circulates back into the room. On a sunny day, the air inside a TAP reaches temperatures of between 90° and 120° F. Unlike a Trombe wall *(page 40)*, however, a TAP has no storage capacity for nighttime heating. Vents must be closed at night to keep cold air out of the house.

Because a TAP can heat only a single room and function only while the sun is out, it is best situated on a room that is used during the day. This may be a child's playroom, a study or a family room. The TAP's peak production will be between the hours of 10 a.m. and 3 p.m.

The TAP design shown opposite can be retrofitted onto a wood-frame house. Like a Trombe wall, it must be installed on a wall that faces within 20° of true south *(pages 8-9)* and has exposure to the sun unobstructed by buildings or trees. Depending upon the amount of such wall space available, the TAP can be constructed one, two, or three panels wide; the example opposite is a double panel. In all cases, the TAP surface area should be no more than one third the floor area of the room you plan to heat.

The panels can be mounted on a first- or second-story wall. One variation on the installation shown here is to build 8-foot-high panels that draw cool air from a room downstairs and vent warm air just above the floor on the second story.

All of the materials used to build a TAP are readily available at lumberyards and building-supply outlets. As on a Trombe wall, it is the size of the glazing material used on the face of the TAP that determines the dimensions of the panels. Although you can use plastic materials made for solar glazing, many experts prefer the double-glazed panes sold as replacement glass for sliding patio doors. Since the edges of such panes are hermetically sealed, they cannot be trimmed to size; rather, the frames of the panels are built to fit around the glass.

To keep the heat that a TAP collects from being absorbed by the wall of the house, the panels are backed by sheets of ⅛-inch-thick foil-faced sheathing made of rigid laminated cardboard. This kind of sheathing is sold in several grades; you will need the most rigid grade.

The absorber plate can be made from corrugated aluminum of the type used for roofs and siding. It is sold in sheets of various lengths and widths, so you should order pieces close in size to the glass panels. Cut the sheets to the length you need by trimming across the ribs with tin snips; to cut along the ribs, score the cutting line with a utility knife and then bend the metal repeatedly until it breaks. A carbide-tipped blade in a circular saw will make the cuts in either direction. Wear eye and ear protectors when you work with a saw.

After the absorber panels have been trimmed to size, use a solution of trisodium phosphate and warm water to wash off the greasy film that coats the aluminum. Rinse the panels, then paint the sides that will face the sun with a flat black heat-resistant paint.

The perimeter frame for the TAP panels can be made with any 2-by-4 lumber. But the stops inside the frame that brace the absorber plates and glass are subject to very high heat. Avoid pine in making those pieces—the fumes that pine emits when it gets hot can fog the glass face of the TAP, making it less efficient. Fir and spruce are good alternatives.

The lumber depicted in the drawings that follow is ¾-inch stock. The 1-inch actual thickness of such stock provides adequate surface area along its edges for seating the glass and aluminum securely. Five 8-foot-long ¾-by-6-inch boards can be rip-cut to provide all the stops you need for a double-panel TAP. These include eight 1¼-inch-wide absorber stops; six 2¾-inch-wide glazing stops for the sides, top and bottom of the TAP; a center glazing strip 2½ inches wide; and a 1¾-inch-wide center mullion. If you have a table saw, you can make these rip cuts yourself; otherwise, have the cuts made for you at the lumberyard. You will also need a supply of 1-by-3s and 1-by-6s for battens that hold the glass in place on the face of the TAP and for trim around the vent openings inside the house.

Among the other materials that you will need is lightweight .020-inch aluminum flashing, both for weatherproofing the top of the TAP frame and for fabricating vent sleeves that ease the flow of air through the panels. For the vent opening inside the house, buy a sheet of the plastic eggcrate grating used to cover ceiling light fixtures. And for the lower vents, buy sheets of 2-mil polyethylene to serve as back-draft dampers when the air in the panel cools. For installing the glass, buy 100 feet of butyl glazing tape ½ inch wide by ³⁄₃₂ inch thick and four neoprene setting blocks ¼ inch thick, ⅝ inch wide and 2 inches long. And finally, buy rubber filler strips to seal off the tops and bottoms of the absorber plates. These are available where you purchase the corrugated aluminum.

Although a TAP is a valuable supplement to your winter heating plant, you will need to shut it down during the warm months by shading it. On some houses, a wide roof overhang may shade the TAP in summer, when the angle of the sun's rays is high. If there is no adequate overhang, you may have to install an awning or rig up a tarpaulin.

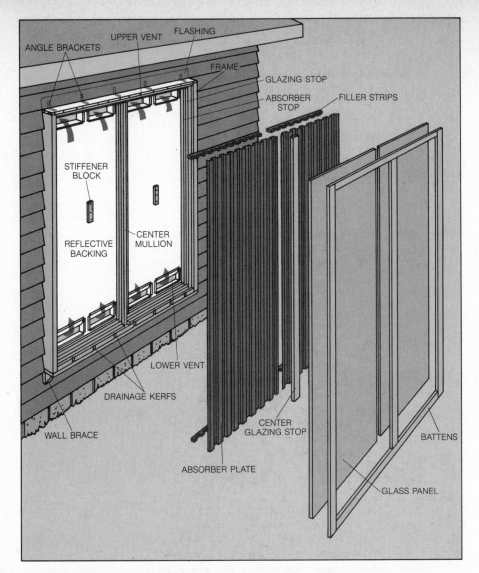

ANGLE BRACKETS
UPPER VENT
FLASHING
FRAME
GLAZING STOP
ABSORBER STOP
FILLER STRIPS
STIFFENER BLOCK
REFLECTIVE BACKING
CENTER MULLION
LOWER VENT
DRAINAGE KERFS
WALL BRACE
ABSORBER PLATE
CENTER GLAZING STOP
BATTENS
GLASS PANEL

**Anatomy of a TAP.** The passive solar collector at left has been installed on the south-facing wall of a wood-frame house. Sunlight passes through double-wall panels of glass and is absorbed by dark-colored aluminum plates. Heat that is conducted through the metal plates is held inside the TAP by a reflective backing that is flush against the sheathing of the house and by rubber filler strips shaped to match the profile of the corrugated aluminum.

Cool air near the floor of the room inside flows into the TAP through vent openings at the bottom of the collector. The vent openings have been cut between the wall studs and through the reflective backing. As the air inside the TAP warms, it rises and reenters the house through vents at the top. During daylight hours, a steady flow of air through the TAP is established by natural convection, which draws a constant supply of cooler air from inside the house.

The siding of the house has been removed around the collector. The 2-by-4 frame is supported by a shelflike wooden wall brace at the bottom and fastened at the top with angle brackets. Inside the frame are wooden stops and a center mullion to hold the metal and glass in position. Stiffener blocks at the centers of the panels keep the metal absorber plates rigid. The front of the glass is held by battens screwed to the edges of the frame. Flashing at the top protects the TAP from runoff rain; drainage kerfs at the bottom of the frame allow accumulated moisture to escape. Thin plastic sheets on the lower vents serve as dampers to prevent reverse drafts.

## Making Thermosiphoning Air Panels

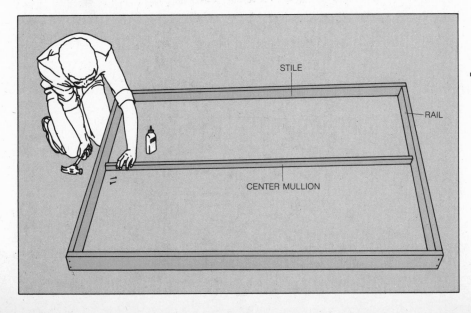

STILE
RAIL
CENTER MULLION

**1** **Joining the frame.** Working on a flat surface, assemble a framework that consists of 2-by-4-inch stiles 3½ inches longer than the height of the glass; 2-by-4-inch rails 1½ inches longer than the combined widths of the two glass panels; and a 1-by-1¾-inch mullion cut to fit between the rails. Join the pieces with yellow glue and 3½-inch common nails, fitting the rails between the ends of the stiles. Install the mullion with a narrow edge facing up, between center points on the rails and flush with the bottom edge.

Make seven evenly spaced pencil marks along the outside edge of one rail. At each mark, make a ¼-inch-deep cut with a circular saw. The kerfs will serve to drain moisture from the TAP.

**2 Attaching the reflective backing.** Turn the frame over so that the saw kerfs face down, then measure diagonally between the corners to be sure that the frame pieces are joined at right angles. The measurements should be equal. Make adjustments, if necessary, by pushing on opposite corners. Spread glue, construction adhesive or caulk on the edges of the frame, then lay sheets of the reflective backing on the adhesive; align the edges of the backing with the edges of the frame. For a double panel, as shown here, you will need to use two sheets of the backing material with their edges adjoining along the center of the mullion. Drive 1-inch roofing nails through the backing at 3-inch intervals around the perimeter of the frame and along the center mullion. Again, flop the entire frame so that the saw kerfs are facing upward.

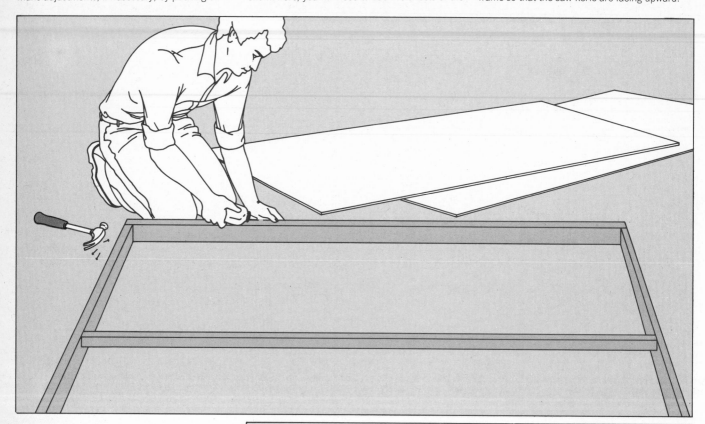

**3 Adding the stops.** Line the inside of the frame rails and stiles with 1-by-2¾-inch glazing stops, seated flush with the backing and secured with yellow glue and 2½-inch common nails. Inside the glazing stops, attach 1-by-1¼-inch absorber stops with their wider edges flush against the glazing stops. A cross section (inset) shows the relative positions of the stops, the frame and the backing. Nail the absorber stops to each side of the center mullion with the wide side flush against the mullion.

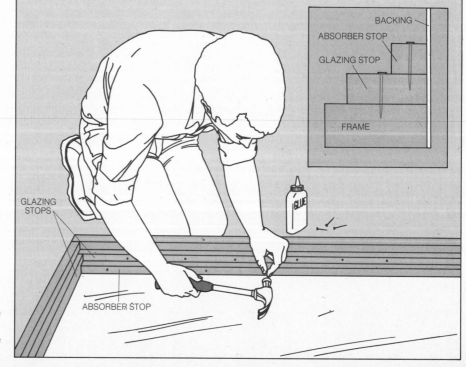

GLAZING STOPS

ABSORBER STOP

BACKING

ABSORBER STOP

GLAZING STOP

FRAME

GLUE

**4** **Plotting the panel location.** On the siding of the house, snap two vertical chalk lines as far apart as the TAP frame is wide. To maximize the number of vents you can build into the panel, situate the chalk lines so that they will be 3½ inches outside the studs at each side. Nailheads in the siding will reveal the stud locations; they will be either 16 or 24 inches apart. After snapping the first chalk line, measure from that line to position the second chalk line.

CHALK LINE

STUD MARKS

**5** **Marking the bottom.** To determine how far down to remove the siding, establish marks outside the house that show the level of the baseboard inside. Begin inside the house by measuring from the top of a window frame to the top of the baseboard. Outside the house, mark the siding at that level after measuring down from the same window.

Hold a carpenter's level at the baseboard mark, and place a second mark where the level crosses a vertical chalk line. From this second mark, measure down 8½ inches (the combined thicknesses of the bottom of the TAP frame and its supports) and use the carpenter's level to mark a horizontal line between the two vertical lines. If there is a slab foundation, position the TAP supports at the baseboard line or above to cross the studs.

Measure 86½ inches up from the bottom line to accommodate the top and bottom of the frame, the supports and the 76-inch-high glass panels used in this example. Make a final mark at that point; this is where the top of the TAP frame will be attached to the wall.

BASEBOARD MARK

**6** **Removing the siding.** With the blade of a circular saw set to the maximum depth of the siding, cut along the lines marked in Step 5. Guide the blade carefully along the outside edges of the lines. If the mark indicating the top of the TAP frame is 3 inches or more from the bottom of the next piece of siding above, cut to the top of the marked siding piece. If the mark is closer than 3 inches, cut through to the top of the next piece of siding as well. This will give you room to install fasteners and flashing above the TAP. Pry off the siding between the cut lines, and set aside the top pieces for later use. If the siding is aluminum, cut it with a carbide blade.

On the exposed sheathing of the house, mark cutting lines for the vent openings. Mark the bottom edge of the lower vents 10½ inches up from the top of the siding still in place. From that line, measure up 72½ inches and mark the top edge of the upper vents. Use a framing square to complete outlines for the vents. Make the vents 6 inches high and as wide as the space between the studs in your wall.

VENT OPENING

**7** **Cutting the exterior vent openings.** Set the circular saw to the depth of the sheathing—usually ½ to ¾ inch. After turning off the power to any wires that may be hidden in the wall, cut out the vents outlined in Step 6. To do this, you will have to repeatedly plunge the saw blade into the surface of the sheathing—a procedure that requires some care. Holding the blade guard open, tilt the saw onto the front of its base until the blade is clear of the wood. Position the blade over the cutting line, then turn on the power and slowly lower the saw until its base rests flat. Grasp the saw firmly as the blade bites into the wood. If you prefer to use a keyhole saw or a saber saw, drill a starter hole at each corner to get the saw blade started.

**8** **Marking the interior vents.** After cutting away the insulation exposed by the exterior vent openings, hammer an awl through the back of the wallboard at the bottom right corner of the right lower vent. Pull out the awl, and punch a second hole at the bottom left corner of the left lower vent. In the same way, mark the wallboard at the top right corner of the right upper vent and the top left corner of the left upper vent.

Inside the house, draw a line connecting the bottom holes and a line between the top holes. With a framing square, mark 3½-inch-high rectangles above the bottom line and below the top line, then cut away the entire length of wallboard within the marked lines. Remove more wallboard at the sides of each hole to expose ¾ inch of the stud on each side of the vent.

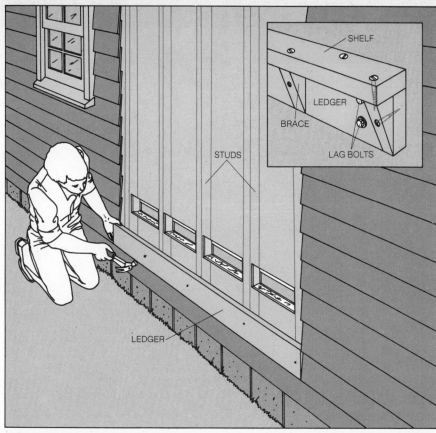

**9** **Building the supports.** To install a supporting ledger, cut a 2-by-6 to the width of the TAP frame and rest it on the siding at the bottom of the opening. Tack the board temporarily in place with 3-inch nails driven partway in. Mark the ledger where it crosses the studs or the band joist, and at 16-inch intervals drill pilot holes for a pair of 3½-inch-long ¼-inch lag bolts. Insert the lag bolts and their washers, and tighten them with a socket wrench. Pull out the nails.

Cut triangular braces from 2-by-6 stock to extend 2 inches out from the top edge of the ledger *(inset)*. Mark positions for these braces at each end of the ledger and between the lag bolt locations. To attach the braces, drill a pilot hole through the front edge of each piece for a 2½-inch No. 10 wood screw. Drive a countersink bit into the pilot holes so that the heads of the wood screws can be seated under the surface of the wood.

Lay a 2-by-4 shelf on the ledger, and screw it to the tops of the braces and into the ledger, again using 2½-inch No. 10 wood screws and countersinking the heads.

**10** **Attaching the frame.** Working with a helper, lift the TAP frame into place atop its shelf, and tack it to the wall with galvanized roofing nails driven through the reflective backing. Secure the top of the frame with 3-inch angle brackets at every stud location. At the bottom of the frame, drive a 3¼-inch nail every 12 inches through the glazing stop into the support shelf. Apply caulk around the frame where it meets the edges of the siding. If the gap is bigger than ½ inch, fill it with fiberglass insulation and then caulk it.

From inside, use an awl to poke holes through the backing of the TAP at the bottom corners of the lower vent openings and the top corners of the upper vents. Go outside and use the holes as guides to mark 6-inch-high vent openings on the backing *(Step 8)*. Cut the vent openings with a utility knife.

**11** **Fabricating vent sleeves.** For each flared vent between the TAP and the room inside, you will need to cut and shape four pieces of aluminum flashing. To make 90° bends in the metal for flanges that will hold the flashing in place, scribe a line with an awl and fold the aluminum over a 2-by-4 nailed to the top of a workbench (*top right*). Wear gloves to protect your hands as you shape the metal.

Begin by covering the studs at the sides of each vent. For these pieces, cut the aluminum 5 inches wide by 6 inches high, and shape ½-inch flanges on the longer edges to wrap around the 2-by-4 studs. Nail one flange to the stud inside the house; nail the other flange to the stud through the reflective backing, using 1-inch galvanized roofing nails. For the vent sides next to the center stud, make one flange to be nailed to the stud inside the house. Bend the flashing around the center absorber stop, and nail it to the side of the stop. Next, make a flat bottom for the lower vents and a flat top for the upper vents. Cut these pieces as wide as the vent openings and 6 inches deep, with a ½-inch flange predrilled at each end for drywall screws that attach the piece to the wallboard inside the house. Nail the outside edges of these pieces to the absorber stops. For the vents next to the center stud, notch the flashing to go around the center absorber stops.

The final pieces of the vent sleeves—the tops of the lower vents and the bottoms of the upper vents—must be curved to ease the flow of air through the TAP. Cut sections of aluminum as wide as the vent openings, and shape a ½-inch flange at the end of each piece. From outside the house, reach through the vent opening and hold the flange against the face of the wallboard inside. With your other hand, bend the flashing in a gentle curve to the rim of the exterior vent opening (*bottom right*). Mark with a felt-tipped pen where the flashing meets the rim; then remove the piece, cut it, and shape a ½-inch flange at that point. For the vents next to the center stud, trim the flashing first so that it can be pulled past the mullion and stops. Put the flashing back in place, and attach it to the wallboard and the backing of the TAP (*inset*). When the sleeves are completed, seal all of the seams in the aluminum with beads of silicone caulk.

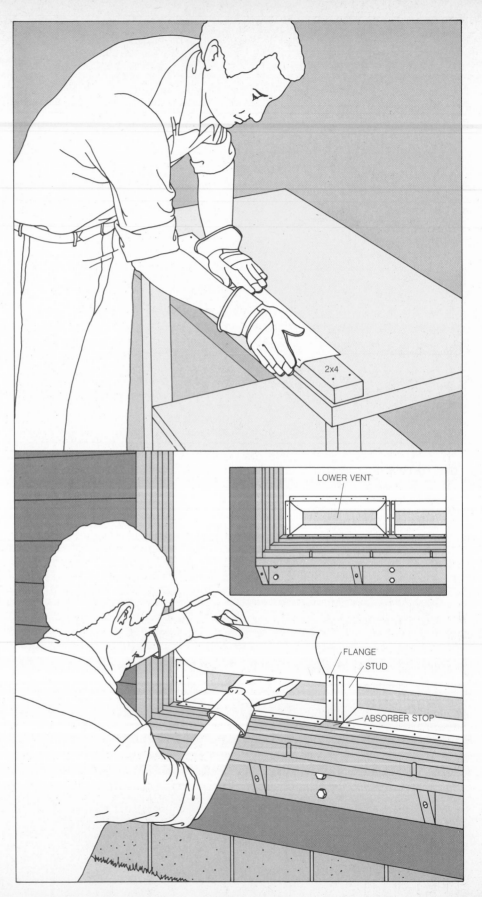

2x4

LOWER VENT

FLANGE

STUD

ABSORBER STOP

ABSORBER PLATE

FILLER STRIP

STIFFENER BLOCK

**12** **Installing the absorber plates.** Cut corrugated aluminum plates to dimensions ½ inch smaller than the spaces between the glazing stops, and with a ¾- to 1-inch flange for nailing on each vertical edge. Clean and paint the plates as described on page 48. On the unpainted sides of the plates, run ¼-inch beads of silicone caulk along the top and bottom edges. Cut rubber filler strips with a utility knife, and press them into the caulk to cement them to the aluminum (*top left*).

Nail a stiffener block—a scrap of wood 1 inch thick and 6 inches long—to the backing of the TAP on each side of the center mullion. Position the blocks so that they will hold the ribbed surfaces of the absorber plates 1 inch away from the backing.

Drill pilot holes at 3-inch intervals around the perimeter of the plates, then apply beads of caulk along those same edges on the unpainted side. Press the plates into place on the absorber stops. Drive 1-inch galvanized roofing nails through the pilot holes into the stops (*bottom left*), then caulk the edges of the plates to make them airtight. Cut the ¾-by-2½-inch center glazing stop to fit between the glazing stops that are already in place on the frame, and nail it to the center mullion. Paint the glazing stops and the nailheads with flat black paint, and touch up scratches in the paint on the absorber plate if necessary.

**13** **Setting in the glazing.** Apply butyl glazing tape to the face of the glazing stops; keep the tape flush with the inner edges of the stops. Lay four ¼-inch-thick, ⅝-inch-wide neoprene blocks on the bottom rail of the TAP frame to cushion the bottom edges of the glass panels. Position two beneath each panel, 16 inches apart. After cleaning the inside of the glass and removing the protective paper from the tape, have a helper assist you in lifting the glass onto the neoprene blocks, then press the glass into the tape. Nail temporary cleats to the edges of the frame to hold the glass until you are ready to install the battens.

Cut three 1-by-3-inch battens as long as the stiles on the TAP frame. Line the front edges of the glass with butyl tape, then attach the vertical battens to the front of the frame, overlapping the tape. Use 1½-inch wood screws to fasten the bars. Screw the center batten to the center glazing stop, overlapping both panels of glass. After removing the cleats, fasten 1-by-3 battens across the top between the vertical battens. For the bottom battens, use 1-by-6s; rip-cut them to 4 inches in width, and bevel the top edges for drainage (*inset*). Caulk between the glass and the inside edges of all the battens.

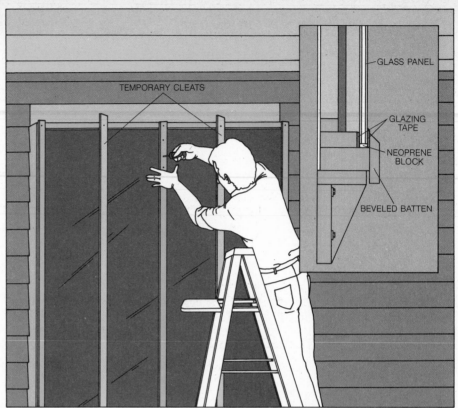

TEMPORARY CLEATS

GLASS PANEL

GLAZING TAPE

NEOPRENE BLOCK

BEVELED BATTEN

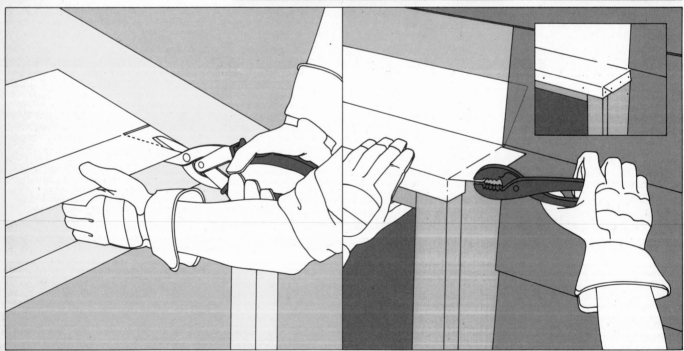

**14** **Weatherproofing the panel.** Cut a piece of flashing 9 inches wide and 3 inches longer than the top of the frame. Scribe a line along the length of the piece, 3 inches in from one of the long edges. Use tin snips to make a 1½-inch cut along the line at each end, then cut diagonally from the corners to make angled notches in the flashing material (*above, left*). Make a 90°

bend on the line. Scribe a line, and make a second 90° bend in the opposite direction, 1½ inches from the other edge of the metal.

Center the flashing along the top of the TAP frame so that it overhangs the sides of the panel by 1½ inches. Press in the front corners, and use pliers to make sharp creases on the flaps that

result (*above, right*). Fold the flaps down against the sides of the TAP frame, and secure them with ¾-inch galvanized nails (*inset*). Drive nails every 10 inches across the front edge of the frame to secure the front of the flashing. Nail the back of the flashing to the sheathing of the house, and then cover it over with the top piece of siding removed in Step 6.

# Grilles and Doors for the Vent Openings

**1** **Installing the grilles.** Cut sections of plastic grating to fit the openings in the wallboard. Cover the edges of the grating with ⅜-inch J molding of the kind used with gypsum wallboard. Set the top vents in place with the wide edge of the molding facing into the room.

For the lower vents, set the grating into the opening and mark the plastic at the edges of the studs it crosses, as well as at the end studs (*below, left*). Pull the grating back out and cut pieces of aluminum flashing 2½ and 1½ inches wide. Fasten the flashing with duct tape to

the thin edge of the J molding so that it will overlap the marks by ½ inch. Then cut sheets of 2-mil polyethylene to cover the grating between the studs. Use duct tape to fasten it to the J molding (*below, right*). Set in the bottom vents with the dampers facing into the TAP.

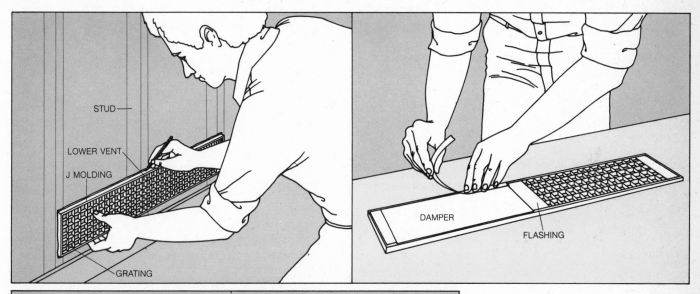

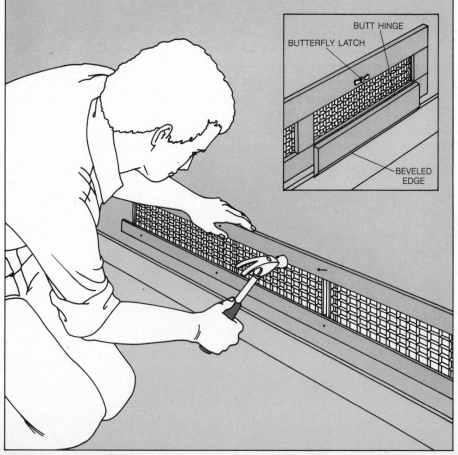

**2** **Adding wood trim and doors.** Cut a pair of 1-by-3s 1½ inches longer than the vent openings. Attach these trim pieces just above and below the vents, using 2-inch finishing nails driven into the studs. The trim should overlap the vent grilles ½ inch to create a vent 2½ inches high. Cut vertical trim pieces to cover the studs. Sink all of the nails with a nail set, and cover the nailheads with wood putty.

Use more 1-by-3s to make a door to cover each vent opening. Plane the top edges of the doors to a slight bevel to allow them to close. Hang the doors, using butt hinges, to the edges of the bottom wood trim; install a butterfly latch on the trim above each door (*inset*). Paint or finish the doors and wood trim to match the other moldings in the room.

# Distributing Solar Heat throughout the House

Passive thermosiphoning solar systems, which rely on the natural circulation that occurs as warm air rises and cool air moves in to replace it, are ideal for heating areas adjoining a south-facing wall (pages 48-57). But if the rooms that need heat are on the north end of the house, an active solar heating system, such as the one described at right, provides an economical and versatile solution.

Although active and passive collectors are similar in appearance, the two systems differ in several important aspects. The active system, instead of thermosiphoning air from the bottom to the top of the collector, uses a blower to pull air horizontally across the collector panels. To heat the large quantities of air in an active system, a larger collector surface area is required: The minimum recommended size is approximately 100 square feet, the maximum 200 square feet. Beyond 200 square feet, so much heat is generated that a special storage system is necessary to conserve it (box, page 63). Finally, active systems use a network of ducts to route heated air through the house and return cold air to the collector for reheating.

An active system offers several advantages over a passive one. A collector mounted on a south-facing bedroom wall, for example, can channel daytime solar heat to a living room or kitchen at the northern end of the house. If the southern house wall is obstructed, the collector can be installed on a south-facing garage or porch and ductwork extended to the living space. And unlike passive systems, active systems can be fully automated, with thermostats that turn the system on when heat is needed and shut it down when collector temperatures drop below the useful range.

The materials and techniques necessary to build the collector are similar to those used for the thermosiphoning air panel illustrated on pages 48-57. The collector is glazed with standard 34-by-76-inch panels of double-insulated glass. Because the characteristics of glazing panels frequently differ, depending on the type of glass and edging used, be sure to follow the manufacturer's recommendations on providing adequate support and allowing room for expansion and contraction as the glass heats and cools.

The heat absorber is made from sheets of black-painted corrugated aluminum, and the wall and stud spaces directly behind the collector are protected from heat and moisture with foil-faced sheathing of rigid cardboard. The collector is fastened to the south-facing wall with angle brackets at the top and with wood blocking underneath.

The only major design difference is the active system's manifolds—two vertical air passages at either end of the collector. The manifolds are created by removing the sheathing and insulation from the stud spaces at the collector's extreme right and left. An intake duct connected to one manifold supplies cool air to the entire collector; a single exhaust duct draws the heated air from the manifold at the opposite end. The construction also differs slightly in the size of the lumber used: A larger frame is used to support the weight of the additional glass in the giant collector and to provide slightly wider air spaces on both sides of the corrugated heat absorber.

The fan that circulates the air is critical to the efficient operation of an active system. The quietest and most effective fans are centrifugal blowers, also known as squirrel-cage fans because of the cylindrical arrangement of their blades. Blowers are sized and rated according to two criteria: the amount of air they move, in cubic feet per minute (cfm); and the resistance to air flow, or static pressure (sp), of the system in which they are used. Static pressure is primarily determined by the depth of the air space behind the absorber plates. The static pressure of the system shown at right is approximately 0.5 inch—a measurement obtained on a special air-pressure gauge used by solar engineers. The necessary cfm rating of a blower is calculated by multiplying the square footage of the collector by 2.5. Thus, a 100-square-foot collector designed like the one at right requires a fan rated at 250 cfm and 0.5 sp, that is, one capable of delivering 250 cubic feet of air per minute at a static pressure of half an inch.

The ductwork must be matched to the air-handling ability of the blower. The dealer who supplies the fan will be able to advise you on the appropriate size of the ducts. They should be run in the most direct possible route from the collector to the room that needs heat. In houses built on slabs, ducts can be run through living areas at floor or ceiling height, then boxed in with wood frames covered with paneling or wallboard. In houses with basements, ducts may be brought out of the collector into the adjoining room, then turned down through the floor into the basement, where they can be hung along ceiling joists. Any ducts that pass through unheated space must be insulated; wrap them in fiberglass batts with the foil facing outward, or purchase factory-insulated ductwork.

The brain of the air-handling system is a differential thermostat, a special 120-volt monitoring device capable of measuring and comparing temperatures at two remote locations. The thermostat body is mounted on a wall near the blower and wired to the blower motor and to the house current as shown on page 62. Low-voltage sensors are installed at the collector's hot-air outlet and in the solar-heated living space, and connected to the thermostat body with bell wire. When the collector temperature rises above the room temperature by a preselected number of degrees—16° F. is a common setting for active systems—the thermostat turns the fan on. When the collector cools to a predetermined temperature, the thermostat shuts the fan off, permitting the collector to reheat or to remain off overnight.

When its fan shuts off, an active system will operate like a passive, thermosiphoning one unless the ducts are blocked in some fashion. During hot, sunny days superfluous warm air from the collector will be thermosiphoned into the living space, overheating it. At night or on cold, cloudy days, the thermosiphoning will work in reverse: Cold air in the collector will drop down to the low, return duct and flow into the living space, while warmed air from that room flows out of the higher supply duct to the collector, where it is chilled. This solar backfiring can be readily thwarted by one-way air valves, called backflow dampers (page 57), or by air registers and grates that open and close manually.

**How air flows in an active system.** The secret of heat generation in an active collector is the horizontal air flow behind the heat-absorber plate: As cool air from the living area enters the collector's intake manifold, it is routed to three separate air channels (*arrows*), spaced 2 feet apart to assure equal distribution of air across the collector. As the air begins its long journey through the channels, it is forced into a sinuous, rippling flow by the corrugations of the heat absorber, which are at right angles to the direction of flow. The resulting turbulence causes the air to come in contact with the entire surface of the absorber, thereby increasing the extraction of heat. At the other end, the air, heated as much as 70° during its passage, is pulled out of the exhaust manifold into the supply duct leading to the living areas requiring heat. A centrifugal blower, controlled by a thermostat with sensors located at the collector outlet and in the heated room, is mounted on the supply duct. Its location on the duct supplying air to the room rather than on the return duct is a deliberate design feature. By pulling instead of pushing air through the collector, the blower makes the air pressure in the collector lower than in the surrounding air. This negative pressure ensures that outside air will be drawn into the collector if any leaks develop. Thus, valuable heated air will be kept from escaping outdoors.

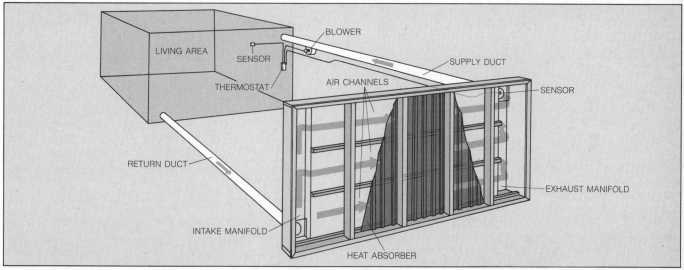

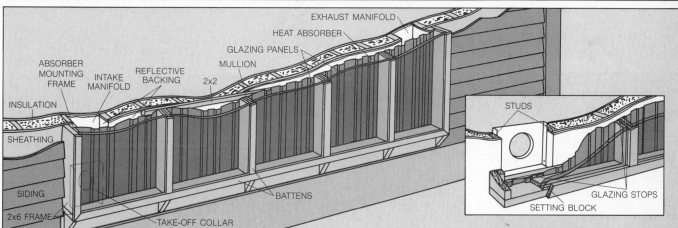

**Building the collector.** A five-panel active collector is framed all around with 2-by-6-inch lumber. The frame is fastened to the wall after the collector's exterior dimensions have been measured and marked and the house siding removed, as shown on pages 51-52. Sheathing and insulation are removed between the pairs of studs at the extreme left and right, to create two vertical air spaces—an intake manifold and an exhaust manifold. A mounting frame for the heat absorber is made by nailing 2-by-2s to the outer frame, with the inner surface of the 2-by-2s flush with the inner surface of the 2-by-6s.

The sheathing, manifolds and exposed surfaces of the absorber mounting frame are covered with foil-faced reflective backing (*page 50*).

Two additional 2-by-2s are then nailed across the foil to create the collector's three horizontal air channels. In each manifold, a hole is cut through to the interior to accommodate the supply and return ducts, and sheet-metal fittings called take-off collars (*inset*) are set into the holes and nailed to the studs on each side of the manifold bays. The absorber plates are screwed to the 2-by-2 mounting frame and the horizontal 2-by-2s, with the edges of adjoining sheets overlapped and caulked.

Vertical mullions—1-by-5-inch lumber lengths—are first rip-cut to 4-inch widths and spaced to provide support where the collector's glazing panels meet. The mullions are then pushed flush against the surface of the ab-

sorber plate and toenailed at top and bottom to the outer 2-by-6 frame. Then glazing stops, fashioned from 1-by-4-inch lumber rip-cut to fit between the absorber and the glazing panels, are nailed to each side of the mullions, to the top and bottom inside surfaces of the 2-by-6 frame between the mullions, and along the inside of the two vertical members at each end of the frame. Panels of double-insulated glass are now installed between mullions, against the glazing stops, using setting blocks to cushion the bottom edges of the glass as shown on page 73. Finally, 1-by-3-inch wood battens are fastened to the outer framing and mullions to secure the glazing panels, all joints between glass and battens are caulked, and the top of the collector is flashed (*page 56*) to channel off rain and snow.

# Fans and Ducts
## for Moving Heated Air

**Hanging ducts from a ceiling.** Map out the most direct practical route for the ducts, minimizing turns and bends in order to maintain the maximum air flow. Then, at 3- to 4-foot intervals along the chosen route, fasten flexible metal hanging straps to the ceiling. While a helper supports one end, lift a prefabricated length of duct into position, making sure that the crimped end points in the same direction as the air flow. Wrap the flexible straps snugly around the duct and secure them with a nut and bolt threaded through the strap perforations.

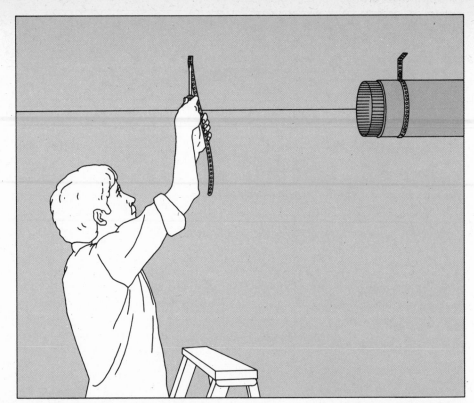

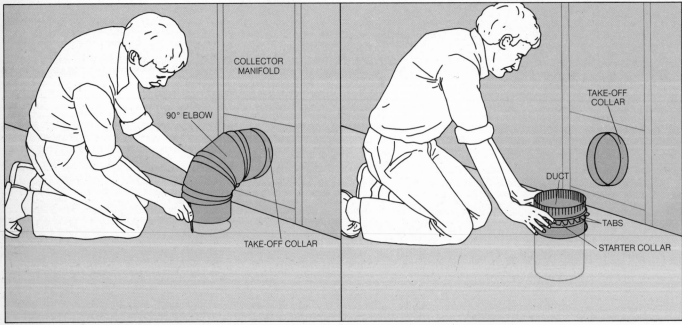

COLLECTOR
MANIFOLD

90° ELBOW

TAKE-OFF COLLAR

TAKE-OFF
COLLAR

DUCT

TABS

STARTER COLLAR

**Routing ductwork through floors.** To turn duct runs down through a floor, you will need three duct fittings matched to the diameter of the main ductwork: a 90° elbow to make the turn; a straight length of duct to pass through the flooring; and a tabbed starter collar to support and secure the other fittings. First, mark the location of the hole by inserting the crimped end of the elbow into the transition fitting protruding from the collector manifold and scribing a circle on the flooring directly underneath the other

end of the elbow. Remove the elbow from the transition fitting and use a saber saw to cut a hole ⅛ inch larger than the marking.

Pry open the seam in the starter collar and bend all the tabs outward at right angles. Wrap the opened collar snugly around the straight length of duct and set the duct into the hole so that the tabs on the collar lie flush with the floor, preventing both fittings from falling through. Adjust the duct within the collar so that its upper end will

meet the lower end of the elbow that extends from the manifold. Mark this position on the duct and remove the duct and collar from the hole. Fasten the collar to the duct with a couple of sheet-metal screws. Reinsert the assembly into the hole and nail every third collar tab to the flooring. Attach the elbow to the duct and manifold, and tape all joints to make them airtight. Continue the duct run underneath the floor by attaching appropriate sections to the protruding end of the straight duct.

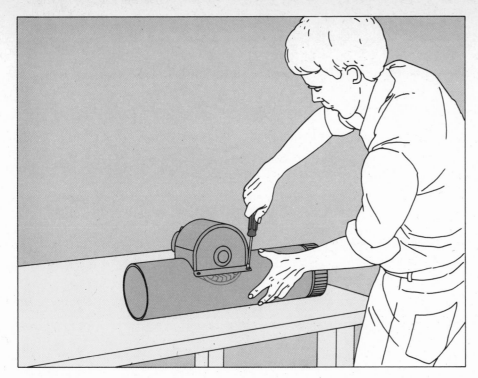

**Mounting a fan in the duct.** While fans of many shapes and configurations are available for active solar heating systems, one of the most convenient and compact units is a blower that attaches directly to the underside of the duct. To mount such a fan, first cut a hole to the fan manufacturer's specifications in a 2-foot length of duct, starting by drilling a small hole and enlarging it with tin snips. Bend the fan's two mounting flanges outward to conform to the shape of the duct and set the fan into the hole. Mark the duct through the two screw holes in each flange, then remove the fan and drill holes at each mark. Reposition the fan and fasten it to the duct with sheet-metal screws driven through the holes in the flange. Attach the duct section to the main ductwork, using hanging straps on either side of the fan to support the extra weight.

## Circulating Heat without Ducts

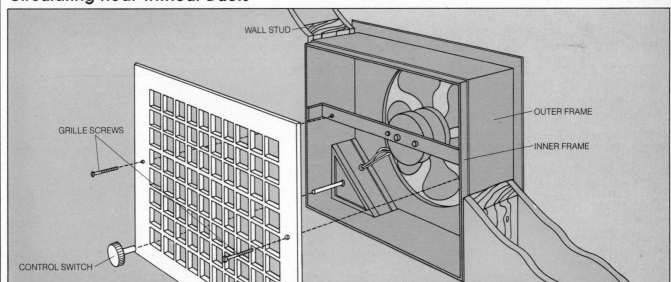

WALL STUD

GRILLE SCREWS

CONTROL SWITCH

OUTER FRAME

INNER FRAME

Circulating air through a network of ducts is not the only way of moving solar heat through the house. A less elaborate method is to use through-the-wall fans. Mounted in a hole between studs, the fans pull air through a grille on one side and blow it into the next room through an opposite grille.

This design lends itself to a variety of solar applications. Installed on an outside wall, the fan can pull air into the living space from an exterior passive collector *(pages 48-57)* or greenhouse *(pages 64-77)*. Installed on an interior wall, it can distribute air from a solar-heated room into an adjoining space. To ensure that cool air flows out as the fan blows warm air in, return registers can be installed near floor level in the same wall as the fan, or a door to the adjoining room can be left open. Both the fans and registers can be adjusted for various wall thicknesses; the model shown has an outer metal frame that slides over the inner frame, changing the unit's depth from 3¾ to 6 inches. The fan may be controlled manually with a switch on its exhaust grille, or it may be wired to a differential thermostat *(page 62)* for automatic operation.

# Wiring the Fan and Thermostat

**Installing the thermostat.** Mount the differential thermostat on a wall near the blower. Then run two-conductor low-voltage cables—one cable for each sensor—from the thermostat to the take-off collar at the collector's exhaust manifold and to the point where the return duct pulls air from the living space. Fasten the cable wires to the screw terminals on the front of the thermostat (*right, top*).

At the collector, drill a screw hole in the top and bottom of the take-off collar. Then drill holes in the copper mounting straps attached to the sensor. Align the holes in the strap and collar and thread a small bolt through each set of holes. Secure the bolts with nuts fastened outside the collar (*bottom*). Drill another hole in the side of the collar and enlarge it so that it will accept a rubber grommet; pull the sensor's wire leads through the grommet and use wire caps to attach them to the thermostat's low-voltage cable. Mount the second sensor in the return duct; connect its leads to the other thermostat cable.

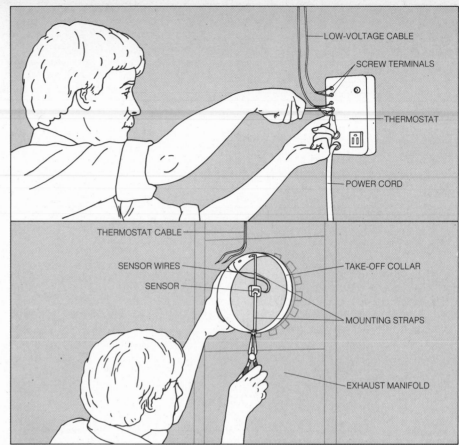

- LOW-VOLTAGE CABLE
- SCREW TERMINALS
- THERMOSTAT
- POWER CORD
- THERMOSTAT CABLE
- SENSOR WIRES
- SENSOR
- TAKE-OFF COLLAR
- MOUNTING STRAPS
- EXHAUST MANIFOLD

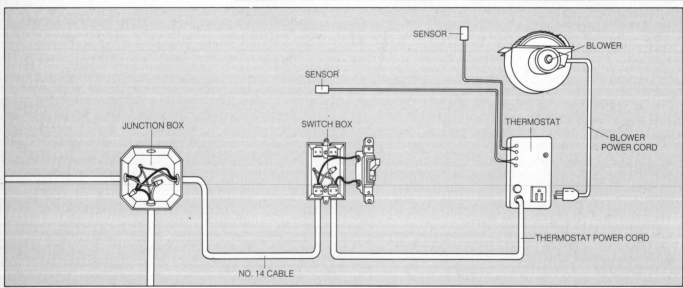

- SENSOR
- SENSOR
- BLOWER
- JUNCTION BOX
- SWITCH BOX
- THERMOSTAT
- BLOWER POWER CORD
- THERMOSTAT POWER CORD
- NO. 14 CABLE

**Bringing power to the blower.** Turn off all power at the junction box or other power source closest to the fan and thermostat. Then mount a switch box on the wall at a convenient location near the thermostat and fan. Run No. 14 plastic-sheathed cable from the junction box to the switch box, securing the ends with cable clamps. In the junction box, connect the new cable's black wire to the existing black wires with a wire cap; similarly, join the new cable's white

and bare ground wires to the existing white and ground wires.

Run the thermostat's power cord to the switch box; if the model does not have a power cable attached, use No. 14 plastic-sheathed cable and connect it to the marked terminals under the thermostat's cover. At the switch box, join the white wires from the power cable and the junction-box cable with a wire cap and join the

ground wires to each other and, with a grounding screw, to the switch box. Connect the black wires to the switch terminals and install the switch and cover plate. Plug in the blower's power cord to the grounded blower outlet on the thermostat front; if the blower cord does not have one, attach a three-pronged grounding plug. Turn on the power to the junction box. You can now use the switch to turn off the solar heating system at night or in summer to prevent overheating.

# Stockpiling Heat in a Rock-filled Bin

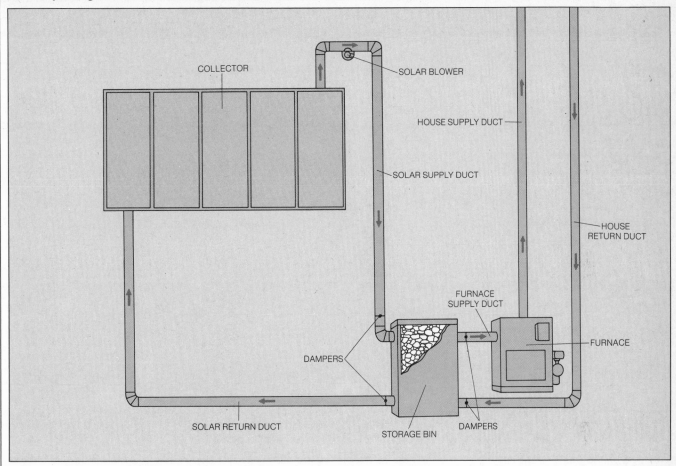

COLLECTOR

SOLAR BLOWER

HOUSE SUPPLY DUCT

SOLAR SUPPLY DUCT

HOUSE RETURN DUCT

FURNACE SUPPLY DUCT

DAMPERS

FURNACE

SOLAR RETURN DUCT

STORAGE BIN

DAMPERS

Solar collectors that are larger than 200 square feet generate so much heat that some form of storage is generally necessary to prevent the house from overheating. On sunny winter days, even the smaller collectors can produce more heat than is immediately needed. If the excess warmth is stored in rocks or other heat-absorbing material, it can then be used during the evening and night hours when the collector is no longer providing heat.

A typical forced-air storage system, shown in simplified form above, utilizes an insulated bin containing 1- to 2-inch rocks that hold the daytime heat. While heat can be stored more compactly in water or phase-change material *(page 32)*, rock storage is usually less expensive, and the containers are less likely to leak heat.

Storage bins can take up a great deal of floor space, because they must hold literally tons of rock—50 to 60 pounds for each square foot of collector sur-

face. In new houses, the bin is frequently integrated into the foundation during construction; in existing houses, it would have to be placed in the basement or on a slab that is strong enough to support the additional weight.

Storage bins can be readily connected to an existing forced-air heating system, which will eliminate the need for separate ducts to distribute the solar heat through the house.

When the sun is shining, a duct-mounted blower draws sun-warmed air through a solar supply duct into the storage bin. After warming the heat-retaining rocks, the air leaves the bin through a solar return duct and is recycled through the collector to pick up more heat. At night, or whenever heat is needed, a return duct from the house brings cool air to the storage bin. Heated by the rocks, the house air flows back into the furnace through a furnace supply duct and from there is pushed by the furnace blower through

the regular house supply duct. When the rocks have given up most of their heat, the furnace kicks in automatically to provide backup heating.

The operation of the blowers is controlled by a differential thermostat with sensors mounted in the collector, in the storage bin and in the living space. Dampers are installed in all the ducts leading into and out of the storage bin; in some of the systems, the dampers are opened and closed manually, while in the more sophisticated installations a series of motorized dampers automatically open and close on command from the thermostat.

This versatile system can even be adapted to provide summer air conditioning in arid climates where humidity is not a problem. During cool summer nights, cold air from the collector is circulated through the storage bin, chilling the rocks. During the heat of the day, house air is cooled as it passes through the cold storage bin.

# The Sun Space: A Solar Collector You Can Live In

A hybrid of the traditional sun porch and the solar collector, a sun space combines the best features of both. As a porch, it provides new living space; as a collector, it works like a greenhouse, trapping solar heat within its transparent walls and roof and storing it in the thermal mass of its insulated masonry floor. The temperature inside a sun space can reach 85° F. or more on a sunny winter day—even when the temperature out of doors is well below the freezing mark. The warmed air, pulled into the house by a through-the-wall fan *(page 61)*, can significantly reduce heating bills.

A sun space is most effective in areas with mild climates—those rated below 3,500 heating-degree days *(pages 8-9)*—but the structure may be modified to suit the harsher weather conditions of other regions. In climates with 3,500 to 5,500 heating-degree days, only part of the roof should be glazed; the upper half should be insulated to help retain as much of the heat of the sun-warmed air as possible. In even colder climates, where more heat would be lost through large areas of glazing than gained through sunlight, the sun space will collect and retain heat most effectively if only the south wall of the structure is glazed.

A sun space must be carefully oriented to take full advantage of the sun's warmth. It should be attached, if possible, to the house wall that faces most directly toward true south *(pages 8-9)*. The orientation may range up to 45° east or west of south, but a variation of less than 15° is ideal.

To prevent prevailing winter winds from robbing a sun space of heat—and to ward off the hot afternoon sun during the summer—the west side wall is frequently closed in, as shown in the example opposite. If the south wall faces more than 15° east of true south, the opposite arrangement is sometimes advantageous: The east wall is insulated and the west wall is glazed.

Any unglazed section of a sun space is constructed like a conventional exterior wall, with 2-by-4 studs sheathed in plywood and covered with siding that matches the house. The comparatively weaker glazed walls are mounted on a sturdy post-and-beam frame of 4-by-4

cedar, redwood or pressure-treated lumber. All of the wooden framing pieces should be treated generously with a wood-preservative stain if you want a natural finish; if the structure is to be painted, use exterior trim paint.

A sun space must be shallow so that sunlight can penetrate its full depth. The side walls should be no more than half as long as the south wall, and the dimensions are further governed by the layout of the walls and doors. The most economical and readily available glazing material for the walls is the type of replacement panel sold for sliding glass doors. These panels come in standard sizes; the design shown here uses panels of 46 by 76 inches. The length of the south wall is determined by the number of panels that are used.

The length of the side walls, one of which contains a sliding glass door to provide access to the outside, is based on the width of the particular door unit that you have chosen. Because doorframe sizes vary slightly, be sure to measure the door you will actually use before laying out the sun space.

The size and slope of the roof are determined largely by the position of the existing second-story windows. The roof must meet the house wall below these windows, preferably at an angle between 30° and 40°. The steeper the pitch, the more sunlight the roof will capture in winter when the sun's arc is low in the sky.

The roof is the most vulnerable part of a sun space. It must admit sunlight, yet resist hailstorms and melting snow without breaking or leaking. Double-wall, ⅝-inch-thick sheets of acrylic, available from suppliers of energy-conservation products or from greenhouse companies, offer virtually the same light transmission and insulating qualities as double-glazed glass panels, yet are far tougher. They are also lighter, making them safer and easier to install. The material comes in a standard 47¼-inch width to fit across rafters spaced 48 inches on center. It can be trimmed to the proper length with a circular saw.

Because acrylic expands and contracts in response to temperature changes more than glass does, the panels are installed

in a special aluminum frame *(page 74)* fitted with silicone-coated gaskets that form a tight yet flexible seal against water leakage.

To keep heat from escaping, a sun space must be airtight. As you build it, bevel the joints and plane the framing members so that they fit snugly. Stuff fiberglass insulation into all spaces behind trim boards or blocking. Use silicone caulk to seal joints, taking extra care where the sun-space frame meets the house wall and along the top edges of horizontal trim—areas that are especially vulnerable to leaks.

The concrete-slab floor of a sun space is a crucial part of its design. To retain the warmth it absorbs from sunlight, the floor is insulated with 2-inch styrene foam boards laid around the foundation walls. For additional thermal mass, you can place containers of water in the sun space, making sure that they are positioned where direct sunlight will strike them; like the floor, the containers collect heat during the day and radiate it at night. Designers of solar heating systems generally recommend 3 to 5 gallons of water for every square foot of glazing on the south wall.

To further increase the light-absorbing qualities of the sun space, the floor and rear wall should be painted a dark color, or the floor may be covered with dark flagstone, slate or quarry tile.

During winter nights, a sun space will lose heat quickly through its extensive glazing. Insulated fabric shades *(pages 19-21)* or pop-in shutter panels *(page 22)* wedged between the posts and rafters at night can be used to conserve some of this heat.

Window shades can be used to block the sun during the summer, and summer shading of the roof is essential to prevent overheating. Nature's shades are another alternative for warm months: Deciduous trees, a row of giant sunflowers or an overhanging trellis of vines will cool the sun space in summer, then die back in winter to allow sunlight through when it is wanted.

Local building codes vary: Your area may require different construction details from those shown here. Have your plans approved before starting.

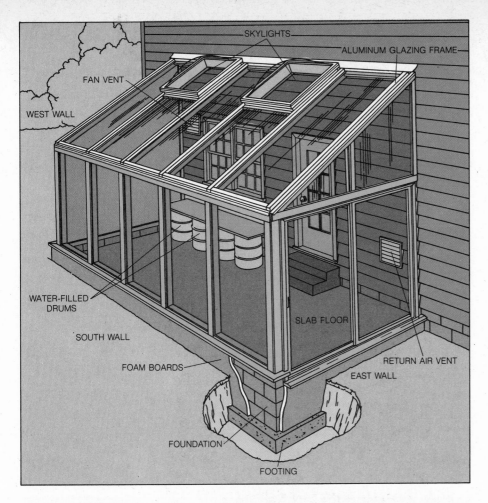

**Anatomy of a sun space.** Attached to the southernmost wall of a house, the 4-by-4 post-and-beam framework of a sun space rests on a concrete-block foundation that supports the slab floor. The outside of the foundation is insulated with 2-inch-thick styrene foam boards to hold in heat absorbed by the slab.

Facing the sunny south, the broad front wall of the sun space consists of 46-by-76-inch double-glazed sliding-door panels mounted between 4-by-4 posts spaced 48 inches on center. The east side wall is fitted with a sliding door. The west side wall is built like an ordinary wall, sheathed in plywood and insulated. The triangular space above the east wall is filled in with double-glazed acrylic panels.

The roof is glazed with acrylic panels, mounted in aluminum frames set on 4-by-4 rafters. Two skylights in the roof open to allow excess heat to escape during warm months. If opening skylights are not used, the roof must be vented. A thermostatically controlled fan like the one on page 61, installed behind a vent in the house wall near the roof peak, pulls warm air into the house; the vent near the floor returns cooler house air to the sun space. Metal drums filled with water add thermal mass to that of the heat-gathering slab and double as a convenient base for a tabletop.

## Building the Foundation

**1 Making square corners.** Mark the house wall where the sun-space foundation will meet it. Nail 8-foot, 1-by-2 marker boards to the wall, with the boards extending 3½ feet inside the corner marks, their top edges level and 16 inches above the intended height of the slab. Make sure the two boards are level with each other. To establish right angles from the house wall, drive one nail into the top edge of each marker board at the corner mark and a second nail 3 feet in from the first. Have a helper hook the end of a tape measure to each nail and cross the two tapes so that the 4-foot mark on the corner-nail tape meets the 5-foot mark on the other tape. Drive a 2-by-2 stake into the ground at this point.

Extend lines beyond the stakes, and drive two more stakes to mark the outer corners of the foundation. Drive a nail into the top of each corner stake. Check the foundation for squareness by measuring diagonally from the nails on the marker boards to those on the corner stakes (inset). If the diagonals are not equal, adjust the position of the corner stakes.

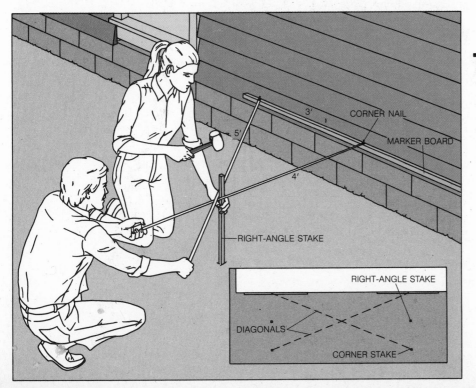

**2** **Setting up batter boards.** To form right angles around the corner stakes, drive three 2-by-4 stakes into the ground about 5 feet away from each stake. Using a water level as a guide, mark the height of the marker boards on the 2-by-4s and nail 1-by-6 horizontal batter boards across them, setting the top edges even with the marks.

To mark the side-wall locations on the batter boards, stretch a string from the corner nail (*page 65, Step 1*) on each marker board to the batter board placed parallel to it, passing the string over the nail in the foundation corner stake (*inset*). Drive a nail in the top edge of the batter board where the string touches it.

To mark the outside edge of the south foundation on the other batter boards, stretch a string parallel to the house wall over the nails in the foundation corner stakes. Drive nails where the string crosses the batter boards. Remove the right-angle stakes and corner stakes.

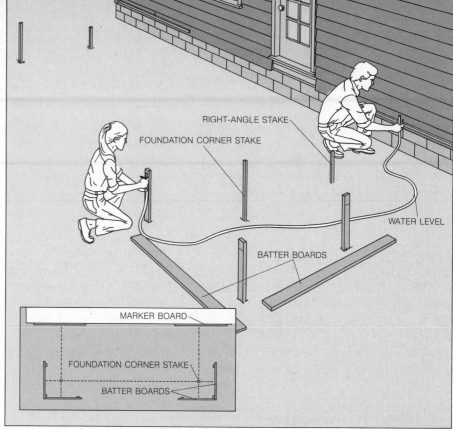

RIGHT-ANGLE STAKE
FOUNDATION CORNER STAKE
WATER LEVEL
BATTER BOARDS

MARKER BOARD
FOUNDATION CORNER STAKE
BATTER BOARDS

**3** **Marking the footing position.** Drive three more nails into each of the four batter boards and the two marker boards, following the pattern and dimensions shown. Those spaced 7⅝ inches apart represent the outside and inside edges of the foundation walls; those placed 4 inches on either side will serve as guidelines for digging a 16-inch-wide concrete footing that will support the foundation.

Outline the inside and outside dimensions of the footing by stretching strings between the appropriate nails on the batter boards and marker boards. With a plumb bob and a squeeze bottle of lime or chalk, mark the string positions on the ground about every 2 feet and connect the marks. Remove the strings and dig a 16-inch-wide footing trench. Measuring from the top of the marker board, dig the trench to a depth that is below the frost line and is a multiple of 8 inches to allow for an 8-inch-deep footing and for the 8-inch concrete blocks that will be used to build the foundation.

To dig to the bottom of a very deep footing, you may need to widen the trench around the outer perimeter. However, the trench for the footing itself must be 16 inches wide and 8 inches deep, with straight walls and a level bottom.

7⅝″
4″
4″
4″
7⅝″
4″
4″
7⅝″

**4** **Leveling the footing trench.** Drive 1-by-2 stakes along the sides of the trench, 3 feet apart in a staggered pattern. Measuring down from the marker board, mark the first stake at the depth for the top of the footing; transfer the mark to the other stakes, using a water level (*Step 2*). If the mark on any stake is less than 8 inches above the trench bottom, dig out the trench farther. Do not fill in any depressions deeper than 8 inches.

Drive a grade peg—a 16-inch length of ½-inch re-inforcing bar—next to each stake and another di-rectly across from it, 3 inches from the walls of the footing trench. Set the top of each peg level with the mark on the stake; remove the stakes.

**5** **Reinforcing the footing.** Lay two lengths of rein-forcing bar in the trench, one along each row of grade pegs. Prop them up with bricks, and wire them to the grade pegs. Where two lengths of re-inforcing bar must be joined, overlap them 15 inches and wire them together. Pour concrete into the trench to the tops of the grade pegs. Spread it with a shovel or a rake, and smooth the top with a float. Cover it loosely with polyethylene sheets and let it cure for 24 hours.

After removing the polyethylene, stretch strings between the nails on the marker boards and bat-ter boards that mark the edges of the foundation wall. Use a plumb bob and lime (*Step 3*) to mark the string positions on the footing, and connect the marks by snapping a chalk line (*inset*). Re-move the strings.

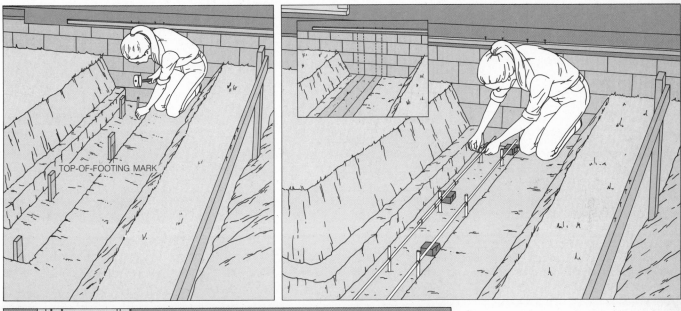

TOP-OF-FOOTING MARK

SHOE BLOCKS

LEAD

**6** **Building the foundation wall.** Lay concrete blocks between the chalk lines on the footing, starting at the corners and ends to provide leads for the wall courses. Run a mason's line between the leads to keep the courses straight. After laying a course of blocks, fill the cores with sand or concrete to add thermal mass to the wall. Brush off the block surfaces so that the mortar of the next course will ad-here. If the wall will be more than three courses high, lay a strip of wire-mesh reinforcement into the mortar bed of every third course.

For the top course, lay L-shaped shoe blocks (*in-set*) to contain the edges of the slab floor. Use solid shoe blocks for the south wall and one side wall, and cored shoe blocks for the side wall that will contain the sliding door. The cores will later hold anchor bolts to secure the sill plate. Plug the gaps left by the corner shoe blocks, using scrap brick and mortar.

# Pouring, Smoothing and Finishing the Slab

**1** **Pouring the slab.** Backfill the trench inside the foundation walls, tamping the dirt level with the bottom of the shoe-block course. Spread 3 inches of gravel over the slab area (*inset*), and level it with a screed made from an 8-foot-long 2-by-4. Cover the gravel with a vapor barrier of 6-mil polyethylene, overlapping the sheets 12 inches at the edges. Run strips of 4-inch expansion-joint filler along the house foundation wall and the shoe blocks. Roll 6-inch wire-mesh reinforcement over the slab area, flatten out the mesh, then slip bricks under it every few feet to elevate it. Pour in concrete, leveling it with the tops of the shoe blocks and filling the shoe-block cores along the side wall. Screed the concrete.

EXPANSION JOINT — WIRE MESH — POLYETHYLENE
SHOE BLOCK — GRAVEL

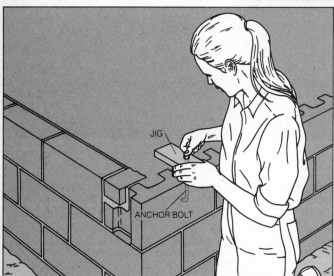

JIG
ANCHOR BOLT

**2** **Installing anchor bolts.** At the wall that will support the sliding door, mark positions for anchor bolts in the wet concrete of the cored shoe blocks. Start 1 foot in from each corner, and space the bolts 3 to 4 feet apart. Make a jig for each bolt by drilling a hole through a scrap of 1-by-4. Insert the bolt through the hole, slip a washer onto the shank and screw on a nut flush with the top of the shank. Sink the bolt into the concrete until the 1-by-4 touches the surface. Leave the jig in place until the concrete sets.

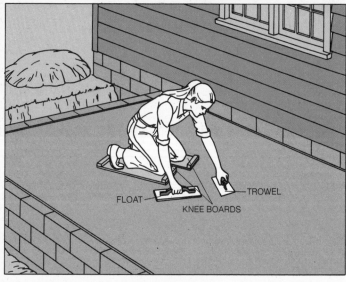

FLOAT
KNEE BOARDS
TROWEL

**3** **Smoothing the slab.** After the concrete in the slab has set for 30 minutes, use a wooden float and a metal trowel to smooth the surface. Supporting yourself on knee boards—plywood rectangles with a 2-by-2 nailed to each end—repeatedly sweep the float over the concrete and follow with the trowel. Work on half of the slab at a time, moving from the middle of the house wall out to the slab edges. When the surface is smooth, spray it lightly with a hose and cover it with 6-mil polyethylene. Let the slab cure for three days.

While the concrete cures, build a knee wall by laying one course of 4-inch-wide cored concrete blocks on the outside perimeter of the south foundation wall and the side wall that will not contain the sliding door. Set an anchor bolt every 4 feet into concrete in a block core (*Step 2*); fill the remaining cores with sand.

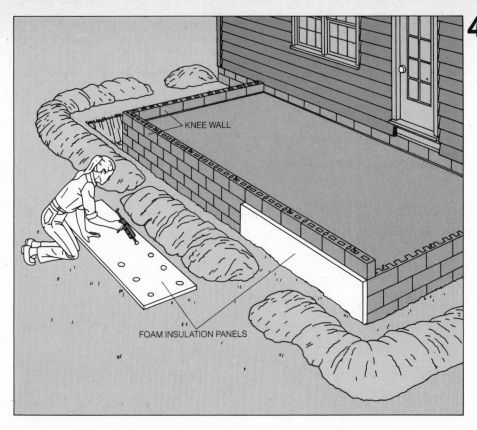

**4** **Insulating the foundation.** Squeeze dabs of construction adhesive at 1-foot intervals in two rows on the back of 2-by-8-foot, 2-inch-thick, styrene foam insulation panels. Press the panels against the foundation wall, with the long bottom edge of each panel flush with the top of the footing. Trim the panels as necessary with a utility knife to cover the entire foundation and knee wall up to the top edge.

Coat the surface of the insulation and the interior of the knee wall with a commercial surface-bonding cement. After the coating has cured—about 24 hours, depending on the manufacturer's instructions for foam—backfill the dirt around the foundation up to grade level, and tamp it firmly.

KNEE WALL

FOAM INSULATION PANELS

## Putting Up the Wall Frames

**1** **Setting the sill plates.** Using a circular saw, cut 2-by-8 sill plates. The south sill should extend 1½ inches beyond the foundation insulation at each end; the side sills should reach from the house to the inside edge of the south wall. For a waterproof drip edge, cut a 30° bevel 2 inches wide down the top surface of each sill. Then cut a lengthwise kerf on the underside of the sill about an inch in from the edge.

To position a sill, set it on top of the anchor bolts with its inside edge even with the inside edge of the blocks below it. Imprint the anchor-bolt locations by striking the sill over each bolt with a mallet, protecting the wood with a piece of scrap. Drill holes for the bolts at these marks, and counterbore ¾-inch-deep holes in the top of the sill to accommodate the anchor-bolt nuts. Set an 8-inch-wide strip of flashing across the top of the blocks, then unroll a strip of sill sealer—a thin strip of insulation—across it, pushing the bolts through it. Set the sill plate in place (*inset*) and tighten the nuts on the anchor bolts.

KERF
SILL
SILL SEALER
FLASHING
KNEE WALL
ANCHOR BOLT
INSULATION

**2** **Erecting the corner posts.** With the aid of a helper, plumb the two outside 4-by-4 corner posts and toenail them to the south sill. Nail temporary 2-by-4 braces to the posts and sills.

Snap vertical chalk lines on the house wall to mark the positions of the inside corner posts. Allow an extra ½ inch of width for the corner post of the covered side wall to accommodate the sheathing of the new wall. Cut away the house siding between the lines with a circular saw. If there are studs in the wall at the post positions, fasten the posts to them with ½-by-6-inch lag bolts. Otherwise, use toggle bolts to anchor them to the wall sheathing, or bolt the posts to nailing blocks installed between the studs. Attach corner posts to masonry walls as shown in Step 2, page 44.

BRACES

**3** **Installing the headers.** Install and brace the post nearest the middle of the south wall. Span the three south wall posts with two 4-by-4 header beams, notched 1¾ inches by 3½ inches at each end for lap joints. Fit the header sections together, and nail them to the posts and to each other at the notches (*inset*). Similarly notch one end of each side-wall header to fit the outside corners of the sun space. Square-cut the opposite ends, and toenail them to the posts at the house wall.

Insert intermediate posts into the south wall every 48 inches on center, and toenail them to the sill and header. To frame the side wall that will be sheathed, toenail 2-by-4 studs every 16 inches to the sill plate and the header.

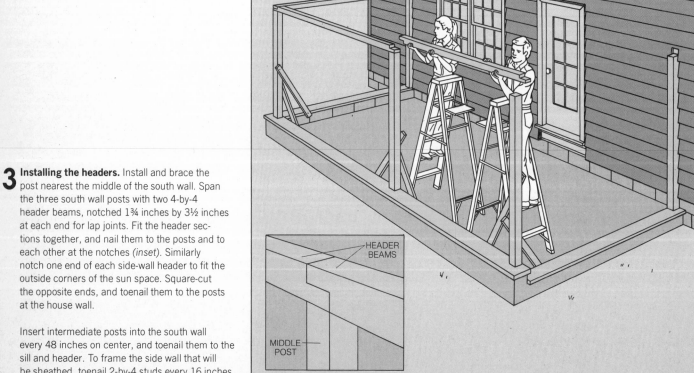

HEADER BEAMS

MIDDLE POST

# Framing the Roof

**1  Attaching the rafter plate.** Cut a 2-by-6 rafter plate as long as the south wall of the sun space, or use two boards butted end to end. Mark the intended location of the roof peak on the house wall by measuring up from each side-wall header, then snapping a level chalk line across the wall. Find the studs in the wall by drilling a hole through the siding just below the chalk line and probing with a wire. While a helper holds its top edge along the chalk line, tack the rafter plate temporarily to the wall. Shim behind the plate as necessary to plumb the face.

Drill a ⅜-inch pilot hole through the middle of the rafter plate into each stud, and insert a ½-by-3-inch lag bolt. For a masonry wall, fasten the rafter plate as in Step 2, page 44.

**2  Making a template for the rafters.** Mark a center line down the face of a 2-by-4. While a helper holds the lower end of the 2-by-4 so that the center line touches the outer corner of the side-wall header, butt the upper end against the house wall with its face against the end of the rafter plate. Hold the top corner of the 2-by-4 level with the top corner of the rafter plate and mark it with a compass: Set one leg against the vertical face of the rafter plate and draw the compass downward so that the other leg scribes a parallel line on the 2-by-4. Take the 2-by-4 down and cut it along the line with a circular saw. Butt the cut end against the front of the rafter plate, and trace the shape of the side-wall header onto the lower end. Extend the lines (*inset*) and cut the 2-by-4 along the lines. Then, using the 2-by-4 as a template, mark and cut 4-by-4s into rafters with a circular saw.

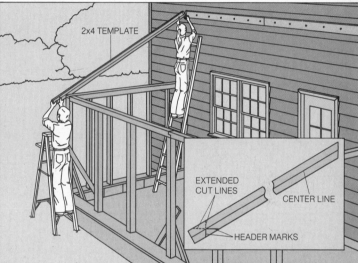

**3  Hanging the rafters.** Nail a 2-by-4 flat against the house wall at each end of the sun space between the rafter plate and the side-wall header. Then install the 4-by-4 rafters, spacing them every 48 inches on center, their lower ends over the south wall posts. Secure each rafter to the rafter plate with two L-shaped metal hangers. Fasten the two end rafters with an L-shaped hanger at the inside and a flat nailing plate over the outside face of the rafter and the end of the rafter plate. Nail the lower end of each rafter to the south wall header.

After installing all the rafters, remove the braces from the corner posts.

**4** **Installing roof blocking.** Toenail a 2-by-4 block between each pair of rafters at the south wall header and the rafter plate to form a continuous flat surface as a base for the roof glazing. Angle each section of blocking so that its face is flush with the top faces of the rafters and its outermost corner is even with the rafter ends.

Toenail two more 2-by-4 blocks between the rafters to support the bottom edges of the skylights. Condensation may run down the underside of the roof glazing. To prevent it from collecting on the interior horizontal surface of the south wall header, fill the V-shaped spaces between the blocking and the header with 2-by-4s beveled to fit, creating a vertical surface that will shed water away from the header.

ROOF BLOCKING

## Enclosing a Side Wall

**Sheathing the wall with plywood.** To enclose a side wall, cut 8-foot sheets of ½-inch exterior sheathing-grade plywood to the height of the wall; nail them to the posts, studs and sill plate. Space the nails every 6 inches along the edges of the plywood and every 12 inches elsewhere. Cover the triangular wall section under the roof with plywood cut to size. Over the sheathing, install siding to match the existing house wall.

Inside the sun space, staple 4-inch batts of fiberglass insulation to the studs as shown on page 30, Step 2. Finish the interior wall with wallboard or plywood.

# Glazing the South Wall

**1** **Building frames for glass panels.** Place a 1-by-4 spacer horizontally across the outside of the south wall header, setting its top edge flush with the tops of the rafter ends. Nail the 1-by-4 to the header every 16 inches. Cut 1-by-2 spacers to fit vertically between the 1-by-4 and the sill; nail them every 12 inches to the center of every post except the corner posts. Nail a 1-by-3 spacer flush with the outer edge of each corner post. Finally, nail a 1-by-4 horizontally to the sill between each pair of posts, aligning its front edge with the fronts of the posts.

Press a length of ¼-by-¼-inch butyl tape—an adhesive gasket that comes in a roll—around all four sides of the frame. Apply the tape to the posts and header ¼ inch in from the post edges, and to the bottom 1-by-4 ½ inch up from the sill *(inset)*. Lay setting blocks—hard rubber pads ⅛ inch thick and 2 inches long—on the sill, 1 foot on either side of each post.

**2** **Installing the glass.** Strip the paper backing off the butyl tape, and lift a glass panel onto the setting blocks. Push the bottom edge of the panel against the tape, then raise the top into the frame, pressing its edges firmly and evenly against the tape. Hold the panel in place with temporary retaining blocks of scrap wood screwed to the frame through predrilled holes.

**3** **Securing the panels with battens.** Apply a strip of butyl tape to the outside face of each glass panel, ¼ inch from the top edge and 3 inches down both sides. Tack 1-by-6 boards across the tops of all the panels, aligning their top edges with the top edge of the 1-by-4 spacer installed in Step 1. Attach the batten permanently with pairs of 3-inch wood screws at 12-inch intervals, 1 inch from its top and 2 inches from its bottom edge. Press strips of butyl tape down the sides of each panel, running the tape 2 inches along the bottom past the corners. Attach a 1-by-4 vertical batten to each post, with 3-inch wood screws located every 12 inches. Finally, apply butyl tape along the bottom edge of each panel, and cover it with 1-inch rounded molding carefully nailed to the sill *(inset)*.

# Installing a Glazed Roof

Most of the components needed for glazing the broad expanse of a sun-space roof are available as patented systems designed for roofing greenhouses and building long transparent walls. One such system, shown at right, consists of framing hardware—aluminum struts called glazing bars that fit together in Erector-set fashion—and double-skinned acrylic panels that slide into them. Sold by structural-plastics distributors, the interlocking parts must be precut and drilled, then assembled on the roof.

The glazing bars come in 25-foot lengths that are cut to size with a circular saw, in a power miter box or with a table saw fitted with an aluminum-cutting blade. The vertical bars that run along the rafters are joined with metal angles to horizontal cross bars, forming a grid that is secured to the roof with wood screws.

The grid supports double-walled, ⅝-inch-thick acrylic glazing panels. Each panel is ribbed, with lengthwise insulating air cavities. The panels are cut to size with a circular saw fitted with an acrylic-cutting blade or fine-toothed plywood blade. To seal the panels against weather or moisture, two types of flexible gaskets—ribbed and wedge shaped—are slipped over flanges on the glazing bars. To drain moisture that condenses in the hollows of the panels, weep holes are drilled through the bottom of the frame.

Because the glazing bars are not designed to house skylights, adapting hardware is used to construct joints between each skylight and the glazing frame. Aluminum angles sold by metal dealers are screwed to the frame. The skylight flashing is set on the angles, and sealed with wedge gaskets. L-shaped aluminum capping angles, attached to the glazing frame with two ⅝-inch sheet-metal screws, cap the ends of the vertical bars.

Before the components of the glazing frame are installed, 6-inch-wide aluminum flashing is nailed along the two end rafters and the bottom edge of the roof, lapping over the edges of the wall trim. When the roof is completed, flashing attached to the house wall is lapped over the upper horizontal glazing frame.

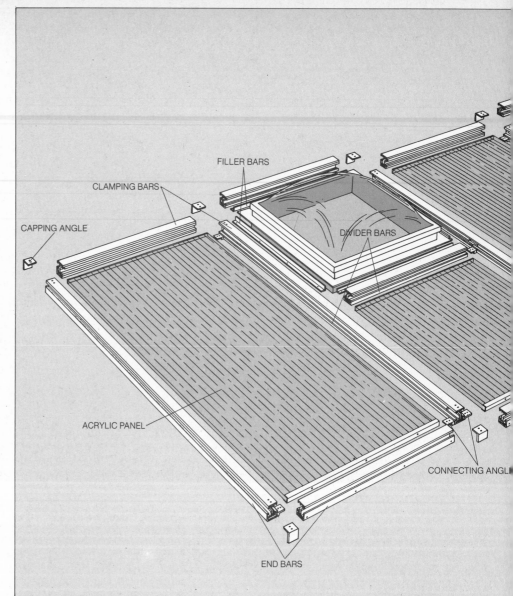

FILLER BARS

CLAMPING BARS

CAPPING ANGLE

DIVIDER BARS

ACRYLIC PANEL

CONNECTING ANGL

END BARS

**Assembling a glazing frame.** Three types of aluminum bars make up a rooftop glazing frame: End bars form the perimeter of the frame; divider bars support abutting acrylic-panel edges; clamping bars fit on top, sandwiching the panel edges against the supporting glazing bars. A wedge-shaped gasket, slipped under the clamping bar, forms a weathertight seal on top of the panel. A ribbed gasket, fitted into a slot in the glazing bar, cushions the panel from below (inset, middle). The frame is attached to the roof with 2½-inch No. 12 Phillips-head wood screws through holes drilled along the center line of the glazing bars. Where bars meet at right angles (inset, top), a connecting angle, attached to the vertical bar with ⅝-inch No. 8 sheet-metal screws, is inserted into a rectangular slot at the end of the horizontal bar. A piece of aluminum called a terminal section covers the open end of each acrylic panel. Weep holes

drilled every foot through the terminal section and the horizontal glazing bars at the bottom of the frame drain condensation from the panels.

At each skylight opening, L-shaped filler bars, ¾ by ¾ by .03 inch (inset, bottom), are attached with ⅝-inch sheet-metal screws to the glazing bars that support the flashing for the skylight. The flashing is held down with clamping bars and wedge gaskets.

To prepare the framing pieces for installation, first cut all of the vertical glazing and clamping bars to the length of the rafters minus 1¼ inches; cut the horizontal glazing bars 45⅝ inches long. (Do not cut the horizontal clamping bars until after the frame has been installed on the roof.) Drill holes 10 inches apart through the center of each glazing bar for the wood screws that will hold the frame to the roof. Cut filler bars

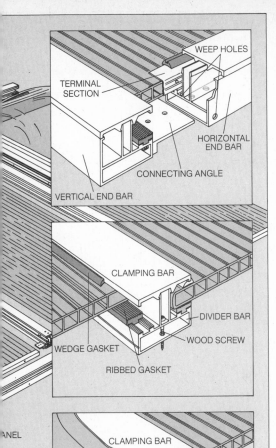

TERMINAL SECTION

WEEP HOLES

HORIZONTAL END BAR

CONNECTING ANGLE

VERTICAL END BAR

CLAMPING BAR

DIVIDER BAR

WOOD SCREW

WEDGE GASKET

RIBBED GASKET

CLAMPING BAR

SKYLIGHT FLASHING

SKYLIGHT CURB

FILLER BAR

ANEL

## Fastening the Frame to the Roof

**1 Mounting the first glazing bar.** Mark a line along the top face of one end rafter 9/16 inch from its outer edge. Align the outer edge of the glazing bar with the mark, its upper end flush with the upper end of the rafter, and clamp the bar to the rafter. Use a variable-speed drill fitted with a No. 2 Phillips-head bit to drive 2½-inch No. 12 wood screws into the rafter through the predrilled holes. If the screws do not go in easily, drill pilot holes. Remove the clamps.

FLASHING

VERTICAL DIVIDER BAR

HORIZONTAL END BAR

for the skylights and drill them, and the framing pieces that will support them, with screw holes spaced 8 inches apart. Attach connecting angles to the ends of each vertical glazing bar and drill four sets of weep holes into each bottom end bar. Drill the ends of the vertical clamping bars and each of the aluminum capping angles with matching screw holes. Use a knife or scissors to cut gaskets for the glazing bars, trimming the gaskets 2 inches shorter than the vertical bars and 2 inches longer than the horizontal bars. Cut the ends of the ribbed gaskets at a 90° angle, and the ends of the wedge gaskets at a 45° angle. Push the ribbed gaskets into their slots in the bars, working from the ends toward the middle.

Cut the acrylic panels 3¼ inches shorter than the vertical bars. Insert the plastic plug strip furnished with each panel into its upper end; cover the lower end with a terminal section.

**2 Squaring the frame.** Slide two horizontal end bars onto the connecting angles of the bar installed in Step 1. Place a divider bar on the adjacent rafter, inserting its connecting angles into the free ends of the horizontal bars to form a four-sided frame. While a helper holds one corner of the frame at a right angle with a carpenter's square, screw the end of the divider bar to the rafter. Similarly, square and secure the other end of the bar, then drive wood screws through the remaining holes in the bar. Drill two holes 1¼ inches apart through each end of the horizontal bars into the connecting angles. Screw the bars to the angles with ⅝-inch sheet-metal screws. Then screw the horizontal bars to the roof.

**3 Installing the skylight.** Install the glazing bars fitted with filler bars (*pages 74-75*) on the roof. Set the curb of the skylight in the framed opening so that the skylight flashing rests on the filler bars. Trim the flashing as necessary with tin snips so that it overlaps the filler bars ⅝ inch on all four sides. Working from inside the sun space, use 2-inch wood screws to attach the skylight curb to the sides of the rafters and the wood blocking.

Continue adding to the glazing frame until you reach the last vertical end bar. To prevent water infiltration, run a bead of silicone caulk around the perimeter of the glazing frame where it meets the roof framing.

**4 Glazing the roof.** Peel back the protective masking from the edges of a glazing panel and position the panel within the first frame of glazing bars. Place a clamping bar on top of the vertical end bar and rap the clamping bar with a rubber mallet to engage the hooks on its underside with the center flanges on the bar. Similarly, install the remaining panels, reaching up from between rafters and through the opened skylights to secure the vertical clamping bars.

When the glazing panels are in place, carefully measure between the vertical clamping bars along each horizontal glazing bar. Cut horizontal clamping bars exactly to these measurements and install them on the horizontal glazing bars.

CURB   FLASHING

PROTECTIVE MASKING

**5 Completing the roof.** Squeeze a wedge gasket under each vertical clamping bar, leaving 1-inch loops of gasket 3 inches from both ends. Compress the loops under the clamping bars—the extra length allows for shrinkage. To reach the top of the roof, hang a ladder fitted with ladder hooks from the skylight frame or roof peak. As you install the gaskets, clean their ends and the joints in the aluminum frame with naphtha, and seal them with silicone caulk.

Screw the predrilled capping angles to the vertical clamping bars. Flash the peak of the roof with a strip of aluminum wide enough to lap over the top horizontal clamping bars, and long enough to extend past the roof six inches at each end. Bend the excess down.

Glaze the triangular section of the side wall with acrylic panels, cut to overlap the frame ¾ inch on each side. Install the panels as you did the south wall glazing, page 73. Then strip the protective masking from all the acrylic panels.

WEDGE GASKET

## Emplacing the Sliding Door

**1** **Positioning the door frame.** Trim the corner posts at the rough opening in the side wall with face boards, bringing their surfaces flush with the surface of the horizontal batten across the top. Assemble the sliding-door frame according to the manufacturer's instructions and insert it into the opening, pressing its attached exterior trim flush against the rough framing. (Sliding doors vary in height, and you may need to nail a 1-by-6 to the rough opening sill plate or the header to reduce the height of the rough opening.) Level the doorframe sill, shimming it as necessary, and fasten it with the screws provided by the manufacturer. Plumb and level the sides and top of the frame, shimming if necessary, and screw them to the rough opening.

**2** **Installing the glass panels.** Seat the stationary panel in its upper and lower channels and push it against the side channel of the doorframe until it locks in place. Then fasten the panel with the hardware provided by the manufacturer. Attach rollers to the bottom of the sliding panel if they are not already installed and insert the panel into its upper and lower channels. Install the door lock and caulk the joints between the door frame and rough opening with butyl caulk.

# Giving Your Water Heater a Sun-powered Start

A solar water-heating system saves energy and reduces bills by cutting the work load of a conventional water heater, thus slowing its fuel consumption.

In warm climates, such systems can supply virtually all of a household's hot-water needs during daylight hours. However, in most instances they work primarily as preheaters.

Solar collector panels trap the sun's heat, which is then transferred to a solar storage tank. City or well-supplied domestic water passes through the tank and is warmed by the stored heat. The warmed water is then drawn into a conventional water heater, which finishes the job, further raising the temperature. An electrical control panel regulates the temperature and the flow of water between the panels and the solar tank.

There are two basic designs for solar water heaters. In what is known as an open-loop system—the batch heater pictured on pages 88-93 is an example—the domestic water supply flows outdoors into a combination solar collector and tank for preheating, then back into the tank of the conventional water heater. Although open-loop systems are efficient—the water is heated directly by the sun, and no energy is lost through heat transfer—the water in them may freeze in cold weather, so they must be drained and shut down during cold months.

The more elaborate closed-loop system (opposite) offers year-round use because the domestic water supply never leaves the house; it is warmed by heat transferred from a second, separate liquid—usually plain water or a mixture of water, antifreeze and anticorrosion chemicals—that circulates in a closed loop between the outdoor collectors and the solar tank. Some closed-loop systems are designed with a drain-back feature, which returns the liquid to the solar tank when the pump is off or when the temperature outside drops below 36° F.

Any solar water-heating system will save some energy. You can make a rough estimation of the cost efficiency of a proposed solar system by following the steps olutlined in the box opposite. To obtain all of the information you will need to make the calculations, contact your utility company to find out the average temperature of your community water supply and ask your local weather service for the average annual number of BTUs of solar energy per square foot in your geographical area.

You will also have to consult a solar-equipment manufacturer or supplier to get a preliminary estimate of the number of collectors you will need—most domestic water systems use between two and four. You will also need to know the square footage of the collectors and their ASHRAE efficiency rating—a standard industry evaluation of collector capacity set by the American Society of Heating, Refrigeration and Air-Conditioning Engineers. The rating—expressed in BTUs—is an average of the amount of hot water an individual collector is capable of producing at various outside temperatures.

In addition to computing cost efficiency, you must study your house to determine whether it can easily accommodate a solar water-heating system. To work efficiently, the collectors—which generally are installed on a roof for better exposure to the sun—must face within 10° of true south and be positioned nearly perpendicular to the sun's rays.

The wooden mounting racks shown on page 80 are designed so that collectors can be fixed at the correct angle on virtually any slope.

The location of the existing water-heater tank is an additional factor to consider when planning an installation. The new solar tank should be located beside the existing water tank, and both should be placed as close as possible to the collectors to minimize the heat loss that is inevitable with long runs of pipe.

If pipes must pass through one or more stories, try to route most of them vertically and keep horizontal runs short. Vertical pipes usually can be threaded through interior partition walls or closets. Running horizontal lines usually entails removing part of the ceiling and cutting holes through joists.

Solar tanks, collectors and control panels designed specifically for domestic water-heating systems are all standard components available through solar-equipment dealers. Collectors vary in size and design. The interior tubes that circulate the heating solution through the collector are made of copper, plastic or galvanized steel; copper is the most common and the most durable material. The glazed portions of the panels are made of glass or plastic.

The number of solar panels that you will need for your system will be determined primarily by the amount of hot water your household requires, taking into consideration such factors as the climate and the latitude where you live and the orientation of your roof. Using calculations similar to those made to compute cost efficiency, a solar-equipment dealer should be able to help you pin down your needs accurately.

All of the additional equipment you will need for the installation is available at plumbing and home-improvement stores. The wooden support racks for the collectors are constructed of 2-by-4, 2-by-2 and 1-by-4 lumber. The plastic spacers used in fastening the racks to the rafters are made from 1/4-inch-thick polyurethane—which is commonly available in sheet form—cut into 2-inch squares easily with a circular saw.

For the plumbing system, purchase Type K copper tubing, following the solar-equipment manufacturer's recommendations to determine the pipe size. You will need enough 90° elbows to run the pipes between the tanks and the collectors, enough pipe strapping to anchor the pipes to a flat surface every 6 feet, and the valves and fittings described opposite. Buy a standard 1/20-horsepower pump for a system with one or two collector panels; for a system with three or more panels or for a two-story house, use a 1/12-horsepower pump.

Buy split-type pipe jackets of fiberglass, elastomer, urethane or isocyanurate to insulate the pipes once they are installed. For outdoor runs use plastic or metal jackets as well, to protect the insulation from moisture and ultraviolet light. Finally, buy a neoprene boot and collar to cover the opening where the pipes pass through the roof and butyl rubber caulking to seal the seam.

Connecting solar components into an existing hot-water system is a straightforward plumbing job. You can readily accomplish the task with standard tools and techniques (page 83).

## A Closed-Loop System with Rooftop Collectors

**A solar water-heating system.** The solar water-heating system below has three main components: a solar loop filled with liquid that transports heat from the rooftop solar collectors to a solar storage tank; a domestic hot-water loop that circulates potable water through piping inside the solar tank to be heated and then passes it into the conventional water heater; an electrical circuit that controls the solar loop.

In the solar loop, sunlight passes into the collector panels to warm a network of narrow tubes soldered to a flat, black collector plate. Liquid passing through the tubes absorbs heat from the metal and carries it through a return line to the solar storage tank in the house. A second pipe, called the feed line, returns the liquid from the tank to the collectors.

In the domestic loop, cold tap water flows into a long coil of tubing inside the solar tank. The sun-warmed liquid surrounding the coil heats the water within. The heated water is fed into the water heater and then flows through hot-water pipes to the rest of the house.

An electrical control panel and two temperature sensors direct circulation in the solar loop. Whenever the first sensor (mounted near the collector) records temperatures higher than the second sensor (mounted at the solar tank), the control panel starts the pump, which circulates the liquid to the collector for heating. The temperature in the conventional water heater is monitored in the normal way by a regular thermostat, which switches on the gas or electric heating system as needed.

Besides the piping and pumps needed to operate the solar water heater, additional plumbing fittings are included to simplify maintenance and repairs. On either side of the pump are gate valves that can be closed to allow the pump and strainer to be cleaned or replaced. Boiler drain valves above and below the solar tank are used to fill and empty the solar loop. On the roof an air vent lets air escape from the loop as it fills with water; a vacuum breaker performs the opposite function, letting air back into the loop when the liquid drains back to the solar tank. Between these fittings is a pressuré-relief valve, which will open should excessive pressure build up in the solar loop. There are two gate valves in the domestic loop that can be closed if it becomes necessary to make repairs to the house plumbing or the solar tank.

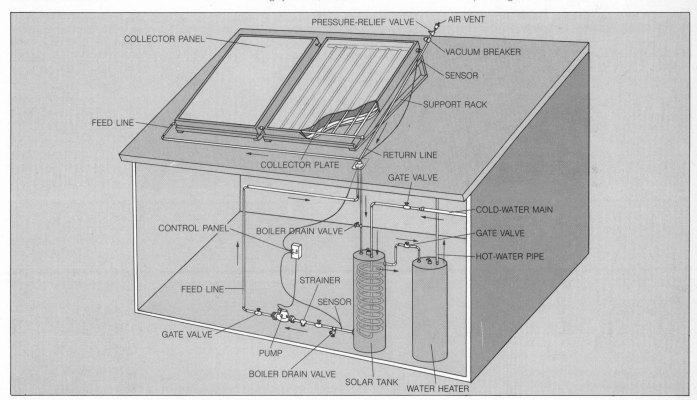

## Computing Cost Efficiency

To estimate the cost efficiency of a solar water-heating system, you must compute both your BTU requirement and the BTU capacity of the system you are considering. First subtract the temperature of the local water supply from the temperature of the hot water in your present system. Multiply the result by 165—the average amount of hot water, in pounds, used by one person per day. Multiply this figure by the number of people living in the house and then by 365 to obtain the household's annual BTU requirement.

Next multiply the average annual BTUs of solar heat per square foot in your area by the number of square feet in the collectors you intend to install. Multiply the result by the collectors' efficiency rating to get the annual BTUs the system will provide. If this figure is at least 40 per cent of the BTU requirement, the system will reduce water-heating bills substantially.

# Making Panel Support Racks

**1** **Calculating the pitch of the rack.** Using a ruler and a carpenter's level, measure the pitch of the roof. First mark the level with a piece of tape 12 inches from one end, and set that end just under the roof's rake board. Hold the level horizontal and set the ruler vertical against the taped mark. Read the distance in inches from the top of the level to the underside of the rake board; convert the number to degrees *(inset)*.

Consult an atlas for the degrees of latitude in your area, and compare this figure with the pitch of the roof. If the difference between the two figures is greater than 10°, you will have to prop up one end of the collector panels, which ideally should face south within 10° of latitude. This additional pitch, angle A, will be used when you are constructing the support rack.

**2** **Constructing the support rack.** To determine the mitered cuts for the two 2-by-4s that form the slanting top of the rack, draw a right angle on graph paper and extend one side of the angle to represent the base of the rack *(inset)*. Intersect the sides of the angle with a line representing the length of the collector panel plus 1½ inches, positioning the line so that angle A equals the pitch calculated in Step 1. Add angle A to angle B (90°), and subtract the total from 180°; the result is angle C.

Mark off the height of the collector on two 2-by-4s, and use a T bevel to mark off angle A and angle C on both boards; miter the boards with a circular saw. Cut two 2-by-4s for the base of the rack and two for the uprights, using the graph-paper drawing to determine their lengths. Bevel the top of the uprights to match angle C. Cut a 2-by-2 front brace equal to the width of the collector panel. Then cut two 1-by-4 back braces equal to the width of the collector panel. Glue and screw all of the pieces together as shown below with 3-inch, flat-head wood screws.

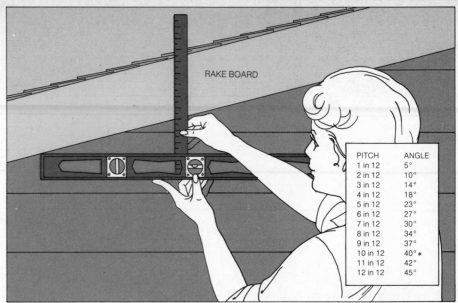

RAKE BOARD

| PITCH | ANGLE |
|---|---|
| 1 in 12 | 5° |
| 2 in 12 | 10° |
| 3 in 12 | 14° |
| 4 in 12 | 18° |
| 5 in 12 | 23° |
| 6 in 12 | 27° |
| 7 in 12 | 30° |
| 8 in 12 | 34° |
| 9 in 12 | 37° |
| 10 in 12 | 40° • |
| 11 in 12 | 42° |
| 12 in 12 | 45° |

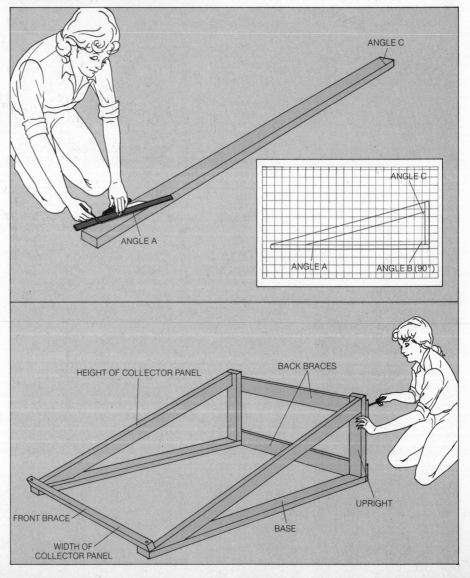

ANGLE C

ANGLE A

ANGLE C

ANGLE A

ANGLE B (90°)

HEIGHT OF COLLECTOR PANEL

BACK BRACES

UPRIGHT

FRONT BRACE

BASE

WIDTH OF COLLECTOR PANEL

## Working Safely on the Roof

Working on a pitched roof presents problems of logistics and safety. The most significant logistical problem—balancing yourself, your tools and your supplies—is easily solved with the aid of two rooftop devices.

A pair of ladder hooks *(below, left)* converts an ordinary extension ladder into a convenient and movable toe-hold—a necessity on any roof with a pitch greater than 4 inches in 12. The hooks, clamped to the two top rungs of the ladder with wing nuts, fit over the roof ridge. A wood block under the

ends of the hooks spreads the weight to prevent damage to shingles.

A pair of adjustable metal brackets supports a level platform for supplies and tools on a roof of asphalt shingles. Each bracket has a shelf that supports a 2-by-10 plank; the shelf is hinged to an upright support. The base of the upright locks into one of a series of slots, so the shelf stays level on a roof of any pitch. The bracket is held to the roof by a steel strap with slots or holes for nails.

To mount the bracket, bend back the bottom edge of a shingle, insert a 2½-

inch nail in a slot or hole, and drive the nail through the shingle below. Remove the bracket by knocking it toward the ridge with a hammer and slipping the strap off the nail. Then pound the nail flush and cover it with roofing cement.

A few simple safety rules apply any time you work on a roof: Always wear rubber-soled shoes to keep from slipping and from damaging roofing material. Try to use cordless electric tools to minimize the danger of tripping over long extension cords. And never work on the roof in rainy weather.

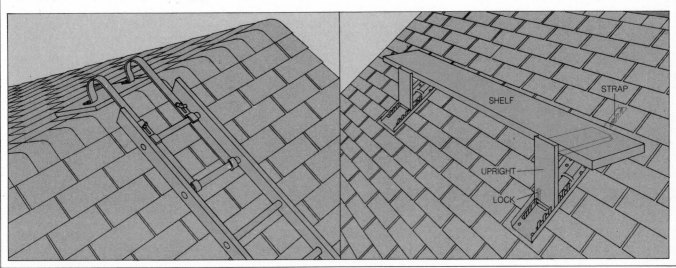

SHELF
STRAP
UPRIGHT
LOCK

## Mounting Collector Panels

COLD WATER INLET

NAIL

**1 Marking the panel locations.** Snap a chalk line on the roof to mark the front of the support rack, extending the line 2 to 3 feet beyond the width of the panel. To keep the line parallel to the eave, measure up from the eave at 3-foot intervals, and make guide marks. Snap a second line at an angle to the first: Make this line slope upward from the planned position of the cold-water inlet at a rate of ¼ inch per foot of total combined collector-panel width. Inside the house, mark off the distance from the first line to the eave, minus the length of the overhang, on the bottom edge of a rafter; drive a 3½-inch nail up through the roof directly beside the rafter. Outside the house, measure ¾ inch from the tip of the nail to mark the center of the rafter. Using this mark as a starting point, mark the center of each rafter along the lower chalk line, spacing the marks 16 or 24 inches apart, depending on the rafter spacing inside the house.

**2** **Mounting the racks on the roof.** Position the support rack with one of the boards cut for the base centered over a rafter, and the front brace aligned with the sloped chalk line; set a ¼-inch-thick plastic spacer beneath the base 8 inches from each end, and drill ⁵⁄₁₆-inch holes through the board and spacers, 1 inch into the rafter. Inject butyl rubber caulking into the holes, and then screw in 6-inch galvanized-steel lag bolts with washers. Drill holes and fasten intermediate bolts at 2-foot intervals along the board, adding a spacer at each hole.

Fasten the opposite base in the same way. If the board does not fall over a rafter, drill from the roof through the board and the plastic spacers, insert a stove bolt through the hole, then cut a 2-by-4 spacer block with a hole drilled through its face and fit the hole over the end of the bolt. Anchor the bolt with a nut and a washer, and nail the spacer to adjacent rafter edges (inset).

Mount additional support racks in the same way, leaving a space of at least 6 inches and no more than 8 inches between racks.

PLASTIC SPACERS

**3** **Hauling the panels onto the roof.** Wrap each panel with a protective covering of heavy brown paper, then loop and tie sturdy rope around each end of the panel, leaving rope ends long enough to reach from the eaves to the ground; knot the rope ends to form a handle. Lean two ladders against the eave, slanting them out from the house a distance equal to one fourth the wall's height. While you stand on the roof and pull the rope handle, have two helpers slowly climb the ladders, pushing the panel up from below.

Rest the panel on the support rack, front edge against the front brace; remove enough of the brown paper wrapping to expose the pipe ends used to plumb the panels into the system, but leave the glazing covered. In the installation shown here, the plumbing is done with copper tubing, as shown opposite.

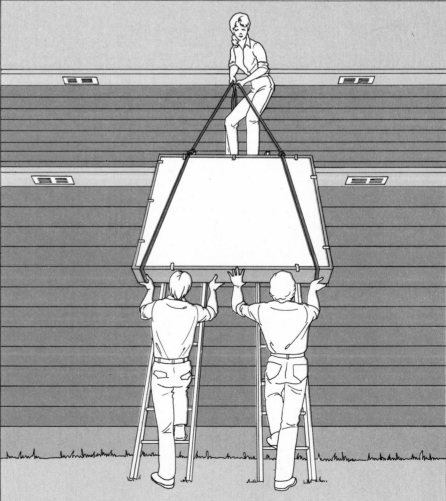

## Joinery Techniques for Copper Plumbing

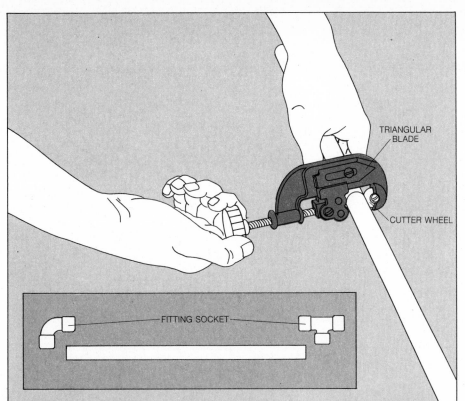

**1** **Cutting copper tubing.** To cut lengths of copper tubing for plumbing the panels, slide a tube cutter onto the tubing at the desired cutting position; at each cut, leave an extra length of tubing equal to the depth of the fitting socket that will slip over the tubing (*inset*). Turn the cutter knob clockwise until the cutter wheel barely bites into the tubing wall, then rotate the cutter once around the tube. Tighten the cutter wheel slightly and rotate it in the opposite direction around the tubing; continue tightening and turning until the tubing is severed. Use the triangular blade attached to the cutter to ream out the burr inside the cut, and file down the ridge left by the cutter on the tubing's outer surface.

TRIANGULAR BLADE

CUTTER WHEEL

FITTING SOCKET

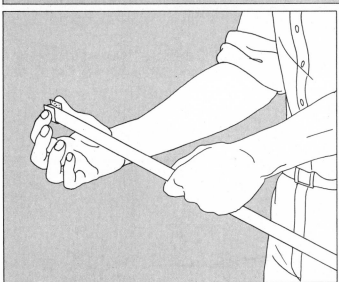

**2** **Cleaning and fluxing the copper.** Clean the copper surfaces to be joined by abrading them lightly until they are shiny—rub pipe ends with emery cloth; scour the inside of fitting sockets with a cylindrical wire brush. Do not touch the surfaces once they are cleaned; greasy fingerprints will prevent the solder from adhering.

Brush a light coat of flux over the cleaned surfaces, assemble the joint and give the tube a twist to distribute the flux evenly.

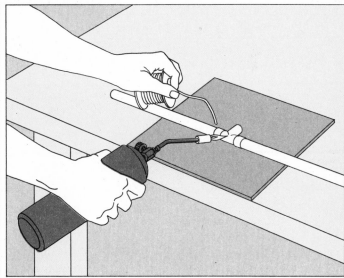

**3** **Sweating the joint.** Lay the fluxed joint on the work surface, protected with a heatproof pad. Light the propane torch, allow it to heat up for a few seconds, then play the flame evenly all around the joint between the pipe and the fitting. Test the heat of the pieces by touching them with a piece of solder; when the solder melts on contact, the joint is ready for sweating. Continue to heat the metal, at the same time touching the solder tip to the joint; but do not allow the flame to touch the solder. Feed solder into the joint until a bead of metal forms around the rim and begins to drip. Then remove the torch and allow the joint to cool.

# Putting Together a Closed-Loop System

**1 Connecting pairs of panels.** Join the pipe ends that protrude from the inner edges of the adjacent collector panels. Use a coupler to make the connection, following the techniques illustrated on page 83. If necessary, push the panels closer together or farther apart to get an exact fit.

Sweat copper caps on two of the four pipe ends along the two outer panel edges, leaving two ends open for hooking up the system—one lower pipe for an inlet and, on the opposite panel edge, one upper pipe for an outlet *(page 79)*.

Attach the panels to the mounting racks with angle brackets at the top and bottom ends.

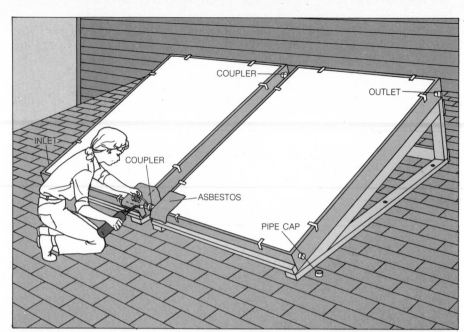

**2 Beginning the feed and return lines.** Mark the point where the feed and return lines will pass through the roof *(page 79)*, locating this opening about 1 inch from a rafter. Then, using copper tubing and 90° elbow fittings, lay out and join the first part of the feed line, sloping ¼ inch per foot from the collector inlet to the mark for the opening in the roof. Join the first section of tubing to the collector's inlet pipe with a reducer *(in-*

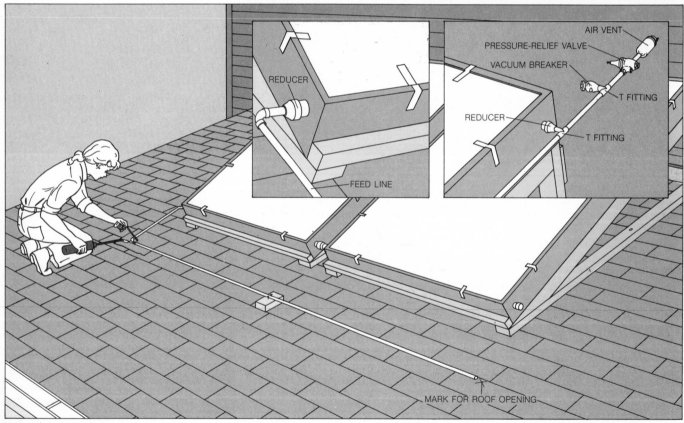

*set, above, left)* that narrows the collector's 1½-inch pipe to a diameter of ¾ inch.

Next, assemble the line leading from the collector outlet to the air vent at the top of the panel *(inset, above, right)*. To start this line, sweat a reducer onto the collector's outlet pipe, and connect a T fitting with a short length of pipe to the reducer. Add enough tubing to the T fitting to

clear the top of the panel. Then join, in order: a T fitting with a vacuum breaker, a 2-inch length of tubing, a pressure-relief valve, another 2-inch length of tubing, and finally an air vent.

At the other opening on the outlet's T fitting, add a length of tubing long enough to lead from the outlet to the mark for the roof opening. This is the first section of the collector's return line.

**3** **Running the lines through the roof.** Drill a ⅜-inch pilot hole through the roof at the point you have marked. Using the hole as a starting point for a saber saw, cut a circular opening, 4½ inches in diameter, through the roof; work from the pilot hole away from the nearby rafter. Position neoprene flashing over the opening; center the boot of the flashing over the opening, and push the upper half of the flashing up under the roof shingles by removing any interfering nails with a pry bar. Fasten the lower half of the flashing over the shingles with a layer of asphalt roofing cement.

Cut two 18-inch lengths of tubing and insulate them with foam *(page 87, Step 9)*. Slide the neoprene boot collar over the tubing and slip the tubing down through the boot, fitting the collar snugly over the boot. Join the top ends of the tubing to the feed and return lines, using elbow fittings. Then, inside the house, anchor both of these pieces of tubing to the nearest rafter, using copper pipe straps.

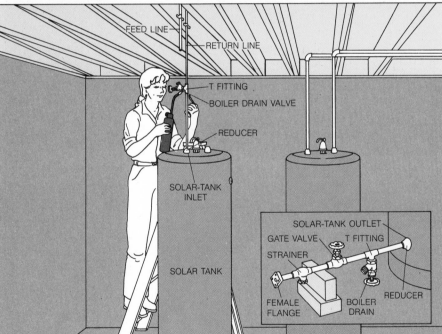

**4** **Plumbing the lines into the solar tank.** Run the return line from the roof down to a point 12 inches above the solar-tank inlet. Join a T fitting to the line, then run tubing the rest of the way down to the tank inlet; there, use a reducer to join it to the inlet. Attach a boiler drain valve to the open leg of the T fitting.

Start the feed line by attaching a reducer to the solar-tank outlet, at the bottom of the tank. Then add, in order: a 10-inch length of tubing, a T fitting with a boiler drain valve, a piece of tubing 3 inches long, a gate valve, another 3-inch length of tubing, a strainer, a third 3-inch length of tubing, and finally the female flange of an electric pump of the appropriate size *(inset)*. As you work away from the tank, support the pipe and fittings on wood blocks.

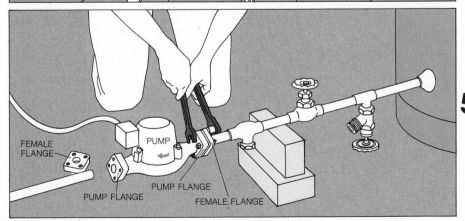

**5** **Completing the feed line.** Set the electric pump so that the arrow on its casing points away from the solar tank, and bolt the pump flange with its fitted washer to the female flange on the feed line. Sweat a second female flange to a 3-inch length of tubing with a gate valve, then bolt this onto the other side of the pump. Add enough tubing and elbows to carry the line up to meet the feed line extending down through the roof.

**6** **Connecting the tanks.** Turn off the cold water that runs to the existing tank and remove the cold-water pipe leading to the tank inlet by breaking the line at the final union above the water heater; use a pipe wrench to twist off the reducer, the pipe and the final elbow that carries the pipe downward. Then, starting with a reducer at the cold-water-feed inlet atop the solar tank, run copper tubing to meet the open end of the cold-water lead; install a gate valve in the horizontal section of new tubing and—if the existing house plumbing is galvanized steel, as shown here—join the new copper line to the old steel line with a dielectric union *(left inset)*. Slip the hex nut and plastic insert of the union over the copper tubing, then sweat the bronze female portion onto the end. Wrap plastic joint tape around the galvanized pipe threads. Twist on the galvanized coupler, the rubber washer, then the hex nut, tightening with a wrench. If the house plumbing is copper, make the connection with a coupler.

Connect the inlet atop the existing tank to the upper outlet on the side of the solar tank with another run of copper tubing *(right inset)*. Install a gate valve on the horizontal pipe and use reducers to make connections at the inlets.

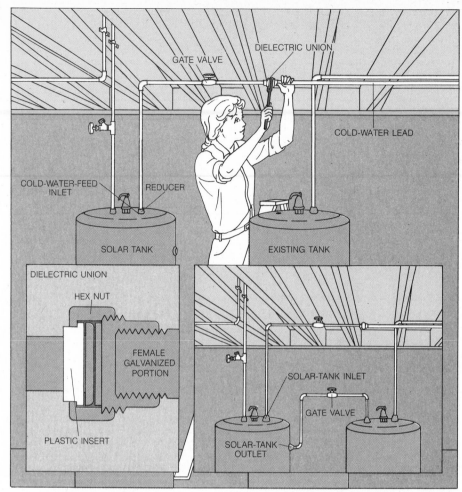

GATE VALVE

DIELECTRIC UNION

COLD-WATER LEAD

COLD-WATER-FEED INLET

REDUCER

SOLAR TANK

EXISTING TANK

DIELECTRIC UNION

HEX NUT

FEMALE GALVANIZED PORTION

PLASTIC INSERT

SOLAR-TANK INLET

GATE VALVE

SOLAR-TANK OUTLET

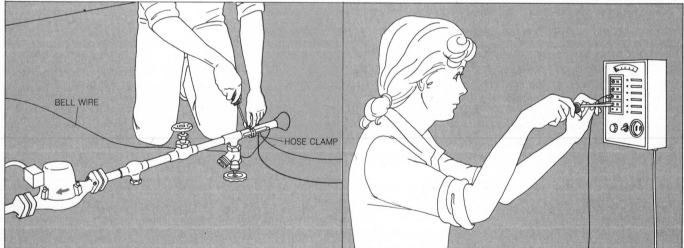

BELL WIRE

HOSE CLAMP

**7** **Wiring the controls.** Mount the control panel for the system on a wall close to the pump. Fasten one of the two sensor probes to the outlet line at the bottom of the solar storage tank *(above, left)*, using a hose clamp to anchor the probe. Tighten the hose clamp until the probe is held firmly against the pipe. Then extend the probe's bell wire to the control panel, and slip the wire's spade connectors under the terminal

screws labeled TANK SENSOR; tighten the screws *(above, right)*.

On the roof, clamp the second sensor probe to the collector's return line, as close to the collector outlet as possible. Run the bell wire down through the roof-vent boot beside the insulated tubing, and seal the top of the roof vent with butyl rubber caulking compound. Then run the bell wire

from the roof down to the control panels, and fasten the spade connectors to the terminals labeled PLATE SENSOR or COLLECTOR. Anchor the collector's bell wire to the return-line tubing with duct tape; anchor the solar tank's bell wire to the wall with cable straps. Plug the power cord for the pump into its socket in the control panel and, with the control panel's power switched off, plug the panel cord into a 120-volt outlet.

**8 Filling the solar loop with water.** Fasten a garden hose to the boiler drain at the top of the solar tank, and open the valve. Close the lower boiler drain valve at the bottom of the solar tank. Open the gate valves on either side of the pump. Run water into the solar loop, stopping when the air vent ceases to hiss. Have a helper check the roof lines for leaks; drain the loop and resolder imperfect joints. When the system is leak-free, let the water drain through the upper boiler drain valve and the hose out of the collectors; close the drain valve. To activate the system, switch on the control panel, starting the pump. Remove the protective paper wrapping from the collector panels.

**9 Insulating the pipes.** Encase all the tubing in the solar loop and the line connecting the solar tank to the existing tank in sections of R-4 foam insulation. Notch the foam with scissors at the corners; tape together the ends of adjacent sections with duct tape. On outdoor lines, be sure the slit in the foam casing faces downward.

To protect outdoor lines further, jacket the foam insulation with sections of split plastic sheathing, again with the slit facing downward. On sloping lines, install the sheathing from the lowest point on the line to the highest, lapping the upper ends over the lower ones. This will keep water from seeping through the joints.

# The Batch Collector: A Simple Solar Water Heater

A solar batch heater is basically just a detour built into the plumbing of the house water-heating system. Cold water headed for the conventional water heater inside the house is routed out through an exterior wall, into a solar collector tank that reclines on a small concrete slab beside the house, and then back through the wall to the water heater. During this jog in its journey, the water is warmed by thermal energy trapped in the solar collector, reducing the amount of work the indoor tank has to do.

Because of its size and shape, the tank-style batch collector absorbs heat less efficiently than the narrow network of thin-walled tubes in the flatplate panel-type collectors of the closed-loop system shown on pages 78-87. And since the batch heater works as part of an open-loop system, where domestic tap water travels directly through the solar collector outside the house, there is a danger that the water may freeze in cold weather. In almost any climate, however, a batch heater works efficiently for enough of the year to substantially cut the cost of water-heating fuel bills; during the coldest months you can simply shut down and drain the system and rely on the existing water heater.

The batch heater is far simpler to build than a closed-loop system, because the collector generally stands on the ground rather than on the roof. And it is a relatively inexpensive system, because you can make the collector yourself and install it with fewer plumbing fixtures and less copper tubing.

The main component of the system—the solar tank—is made of galvanized steel and generally holds 30 to 40 gallons of water. You can buy a tank designed especially for use as a batch collector from a plumbing-equipment dealer, but often a discarded water-heater tank will serve just as well; it needs only a few simple plumbing alterations to accommodate a cold-water inlet that may be located either on the side of the tank near the bottom or on the bottom end, as shown opposite.

If you find a used tank at a junk or scrap-metal dealer, you will have to cut away the outer sheet-metal housing and check the inner tank to be sure there are no cracks, holes, split seams or rusted spots; check for leaks by blocking all but one of the tank openings and filling the tank with water. Then clean the tank thoroughly. First, sand rough spots on the surface with medium-grit sandpaper, and then wash the entire tank with soap and water; use denatured alcohol on grease spots that soap will not remove. Finally, cap off all but the inlet and outlet openings on the tank.

To turn the tank into a solar collector, you must increase its capacity for heat absorption by darkening its surface. You can paint the tank with an exterior flat black paint or apply a highly absorptive and nonreflective black coating made of nickel chromium; the latter comes in rolls with an adhesive backing that makes it quick and easy to apply. The cylindrical surface of the tank shown on the opposite page is covered with such an adhesive coating; the top and bottom of the tank—difficult to fit exactly—are painted black. The paint is available from any paint dealer, and the rolls of adhesive tank coating can be bought from solar-equipment dealers.

A thin plastic film called Mylar, aluminized to make it reflective, is used to cover the walls of the wood cradle that supports the tank, further increasing its heat absorption. Other types of plastic film are used to make the three layers of glazing that cover the collector and trap the sun's rays inside the tank compartment. The two inner layers of glazing are made from a Teflon film 1 mil thick or, if Teflon film is unavailable, from any of the clear plastic films sold in rolls in hardware stores. The single exterior layer is made of a 7-mil acrylic-polyester laminate called Flexigard but any of the rigid plastics listed in the chart on page 17 are suitable, as is tempered glass.

The framework that supports the solar tank and its various reflective and transparent surfaces is made from pressure-treated 2-by-4s and 2-by-2s along with several sheets of ¼-inch hardboard and

⅜-inch exterior-grade plywood. Insulation is provided by 3½-inch-thick glass-fiber batts with plastic or aluminum vapor barriers. The glazed panels are anchored on top of the frame with silicone caulk and wood screws; the edges of the panels are protected from moisture by strips of aluminum flashing.

The plumbing system consists of Type K ¾-inch copper tubing with T fittings and 90° elbows; three gate valves; a boiler drain valve; a vacuum breaker and a pressure-relief valve. The interior hot-water line is insulated with split foam jackets, all exterior lines with foam and plastic jackets (page 87). The openings in the house wall for the cold- and hot-water lines are caulked with butyl rubber compound. All of this equipment is available at a plumbing-supply store, but before you buy it, plan the location of the collector so that you can determine how much copper tubing to purchase.

It is best to build the collector as close as possible to the inside water heater, in order to reduce the heat loss that is inevitable when water must travel through long pipe runs. But like any solar collector, the batch heater must face within 10° of true south to work effectively (pages 8-9). A south-facing wall is the logical location, but if you must choose a wall that faces in another direction in order to keep the collector close to the existing water heater, position the collector far enough from the house wall so that it can face south and remain unshaded for most of the day.

In calculating the correct angle of inclination for the tank, use the instructions on page 80, Step 1. Plan to set the tank tilted on end as shown opposite, rather than on its side. This creates a more distinct separation between the cold water entering at the bottom of the tank and the heated water leaving at the top, allowing the system to work better.

The construction of the batch heater requires basic carpentry skills and the plumbing techniques illustrated on page 83. You must also pour a small concrete slab to support the weight—up to 500 pounds—of the water-filled solar tank.

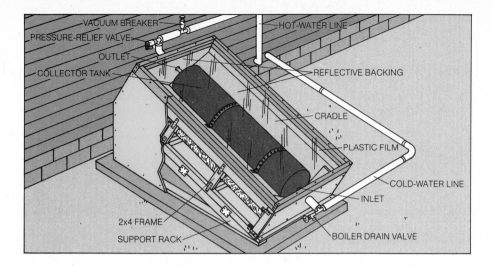

**Anatomy of a solar batch heater.** Fashioned from a discarded water-heater tank, this solar collector rests in a wood cradle supported by a rack of 2-by-4s. The structure is sheathed in plywood and stands on a 4-inch concrete slab, which bears its considerable weight.

During operation of the system, the sun's rays are trapped in the cradle compartment by three layers of transparent plastic film. They bounce off the reflective backing that coats the cradle and penetrate the tank. Domestic cold water, entering the tank through the inlet at the bottom, is heated by the trapped thermal energy and then routed through the outlet at the top to the conventional water heater.

## Supporting the Solar Tank

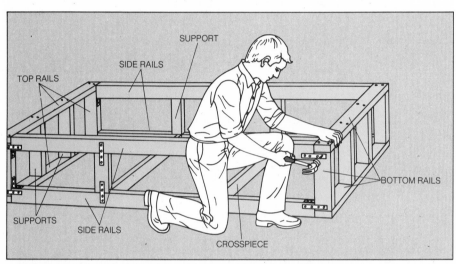

**1 Constructing the tank-frame exterior.** Nail 2-by-4s together to form a rectangular frame. The distance between the rails on opposite sides of the frame should be 12 inches greater than the diameter of the tank; the distance between the upper and lower side rails on the same side of the frame should be 12 inches greater than the tank's diameter. The distance between the top and bottom rails should be 12 inches greater than the height of the tank.

Install supports to divide the top, the bottom and each side into thirds; divide the underside of the frame into thirds with two crosspieces. Cut the pieces and join them as shown.

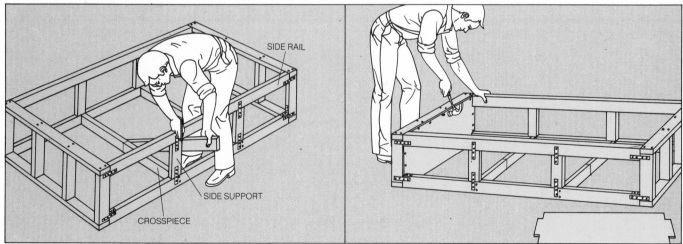

**2 Cutting diagonal supports.** Mark center points on the faces of each crosspiece, then set a 2-by-4 diagonally on edge, its lower end aligned with a center mark and its upper end against both the edge of a side support and the bottom of a side rail. Mark the upper end of the 2-by-4 with a vertical line along its face; use the outer edge of the side support as a guide (*above, left*). Cut the board, then trim the corner off the other end to make the ends parallel. Cut seven more 2-by-4s to match, using the first board as a template.

Cut rectangles of ¼-inch hardboard to match the outer dimensions of the top and bottom of the frame. Then cut notches 1½ inches deep and 3½ inches long at the four corners of each rectangle, and nail the pieces to the inside edges of the top and bottom rails (*above, right*).

**3** **Installing the diagonal supports.** Position pairs of diagonal supports end to end to form Vs extending from the center of each crosspiece to the edges of opposite side supports; nail the 2-by-4s to the crosspieces and to the side supports. At the top and bottom of the frame, nail through the diagonal supports and the hardboard into the frame pieces. Then, along the top edges of both side rails, mark the location of each diagonal support.

Cut two panels of ¼-inch hardboard to form a V-shaped tank cradle that will rest atop the diagonal supports; nail the panels to the top edges of the supports. Then cut two 2-by-6 boards 12 inches long, and lay them in a V inside the tank cradle, flat against the sides, their edges butted against the planned bottom end of the frame (*inset*). Anchor the boards by driving wood screws through their faces and the hardboard, into the diagonal supports below.

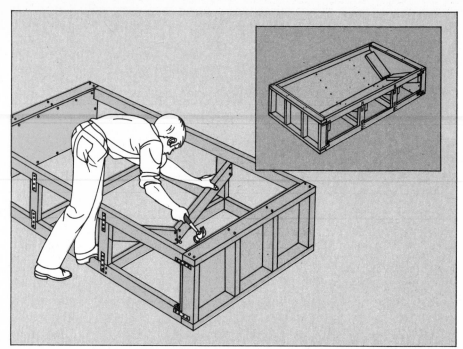

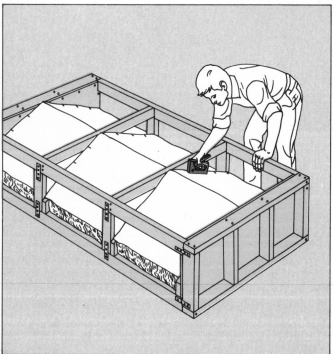

**4** **Insulating the frame.** With the frame face down, trim fiberglass insulation to fit over the hardboard cradle between the diagonal supports. Lay the batts over the cradle, vapor barrier up, and push the fiberglass in between the hardboard and the side rails; staple the vapor-barrier edges at 6-inch intervals to the edges of the diagonals. Fit rectangular pieces of insulation between the supports at the top and bottom of the frame, vapor barriers facing out, and staple. Cut ⅜-inch exterior-grade plywood to fit over the ends and underside of the frame; nail the plywood to the frame.

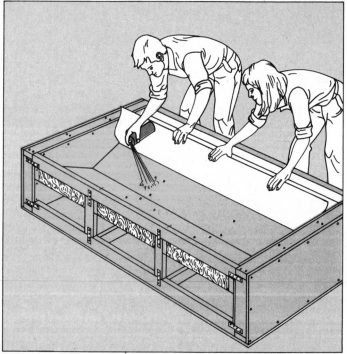

**5** **Installing the reflective backing.** Cut sheets of reflective backing (*page 88*) to fit the sloped sides of the tank cradle, the triangular sections of hardboard at the top and bottom of the cradle and the 2-by-6 supports at the bottom end of the cradle. Lay each sheet in place and, while a helper holds the edge, roll the sheet back and spray adhesive over the wood with long, even strokes. Lay the sheet back down and wipe the surface with a soft, clean rag, working from the center outward, to stick the backing down and smooth out wrinkles and air pockets.

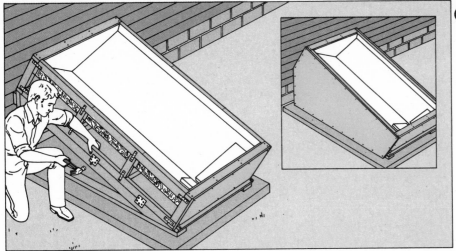

**6** **Building a support rack for the frame.** Determine the correct angle of inclination for the collector according to your latitude *(page 80, Step 1)* and construct an angled support rack for the tank frame, as described on page 80, Step 2. Set the rack on its concrete slab, then anchor the tank frame to the rack, using steel truss plates and nails.

While a helper holds a large sheet of ⅜-inch exterior-grade plywood against the side of the assembly, trace the outline of the structure's side on the plywood. Cut around the outline, and nail the resulting plywood panel to the rack and frame to enclose the side of the structure. Mark and cut a similar panel for the other side, and nail it in place *(inset)*.

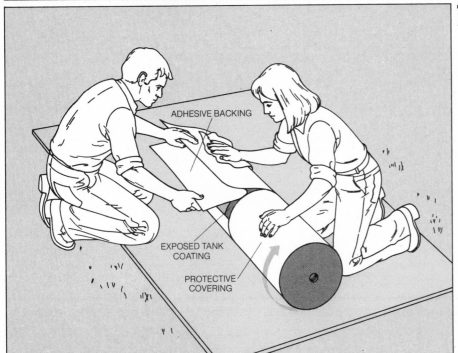

ADHESIVE BACKING

EXPOSED TANK COATING

PROTECTIVE COVERING

**7** **Coating the tank.** Paint the ends of the tank with flat black exterior paint and let the paint dry. Then, working on a clean, flat surface, cut a strip of black tank coating *(page 88)* 1 inch longer than the circumference of the tank, peel back one edge of the adhesive backing and press the exposed edge against the tank, aligning the end of the coating strip with one end of the tank. Then peel the backing off slowly and evenly as a helper rolls the tank away from you and presses the coating against the tank with a soft rag, rubbing gently from the middle of the strip outward to smooth away wrinkles and air pockets.

After the first strip of coating has been applied, peel away 1 inch of protective covering from its inner edge and cut off the strip of covering with a utility knife; do not damage the coating underneath. Then cut and apply a second strip of coating, overlapping the exposed edge of the first strip. Continue until the tank is completely covered, trimming the last strip of coating as necessary for an exact fit. Pull all of the protective covering off the coating.

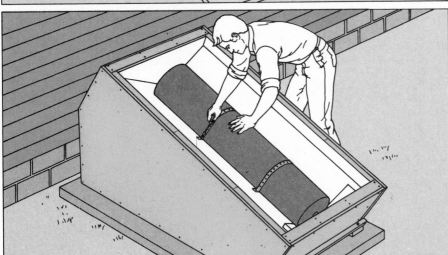

**8** **Installing the tank.** With a helper, set the tank in the cradle with its lower end against the 2-by-6 boards at the bottom; be careful not to slide the tank over the easily torn reflective backing in the cradle. Anchor the tank in place with pipe straps screwed to the diagonal supports below the cradle; use the marks on the frame side rails to locate the supports.

# Hooking Up to the Existing Water Heater

**1** **Running copper tubing from the tank.** At each end of the tank frame and in line with the tank inlet and outlet, drill a 1-inch hole through the layers of plywood, hardboard and insulation. At the bottom end, screw a ¾-inch reducer into the cold-water inlet and then, protecting the wood of the frame with asbestos, use the pipe-sweating techniques illustrated on page 83 to extend a length of pipe 4 or 5

inches beyond the bottom of the frame; sweat a T fitting onto the end of the pipe. Sweat a boiler drain valve onto one leg of the T; to the other leg sweat a run of pipe leading to the spot where the pipe will enter the house.

To the hot-water outlet at the top of the tank add a reducer, a short length of pipe and a T fitting. Sweat a pressure-relief valve to one leg of

the T and to the other leg a short length of pipe, a T with a vacuum breaker and then a run of pipe to the planned opening in the house wall.

Drill holes through the house wall at least 2 feet above grade. Insulate the pipes with foam and plastic jackets *(page 87, Step 9)*, and run them into the house; caulk around the holes with butyl rubber caulking compound.

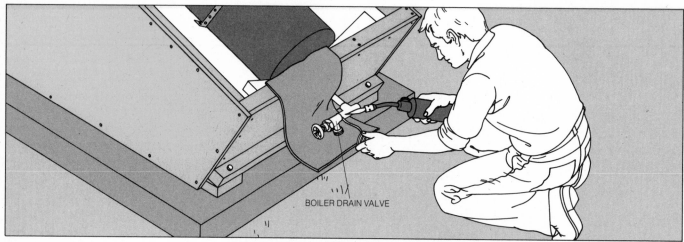

BOILER DRAIN VALVE

**2** **Connecting at the water heater.** Switch off the water heater, shut off the cold water, then remove the final elbow and the attached vertical run of pipe leading to the tank's cold-water inlet. Sweat a T fitting onto the line; to the vertical leg of the T sweat a short length of pipe with a gate valve, another T fitting and a pipe with a ¾-inch reducer to connect at the tank inlet. To the free leg of the upper T sweat a run of pipe, with a gate valve, leading to the collector's cold-water line. To the lower T sweat a run of pipe, with a gate valve, that leads to the collector's hot-water line *(inset)*. Insulate the lines.

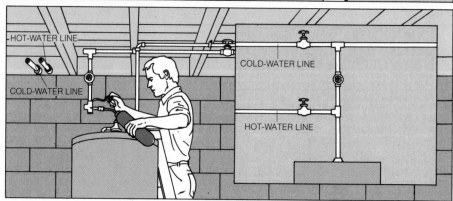

HOT-WATER LINE

COLD-WATER LINE

COLD-WATER LINE

HOT-WATER LINE

# Glazing the Solar Collector

**1** **Making the panel frames.** Construct two frames of unwarped 2-by-2 lumber; their outer dimensions should match the outer dimensions of the front face of the tank-support assembly, including the edges of the plywood that covers the ends and sides of the structure. Cut lap joints *(inset)* for the corners of each frame, and fasten the joints with 1-inch flat-head wood screws, countersinking the screwheads. Screw angle irons to all of the inside corners; then paint the frames and the outside of the support structure if desired. Allow the paint to dry completely; the 2-by-2 frames should be laid flat while drying so that they will not warp.

**2** **Stapling film to the panel frames.** Working on a clean, flat surface, lay a 2-by-2 frame atop a sheet of Teflon film *(page 88)*, and cut a rectangle of film 3 inches longer and wider than the frame. Make a V-shaped cut from each corner of the film into the corner of the frame; then, working on the longer sides first, fold the edge of the film up over one side of the frame, and staple the film to the wood at 3-inch intervals. Pull the film taut at the opposite side of the frame, and staple it in the same way. Staple the film

over the shorter ends of the frame. Then flip the frame over, lay it atop a second sheet of film and repeat the same procedure.

Lay the second 2-by-2 frame atop a sheet of acrylic-polyester laminate *(page 88)*, acrylic side up. Cut and fasten it as before; leave the opposite face of the frame uncovered.

Apply a continuous bead of silicone caulk around the top edge of the tank frame. Then, with a

helper, lay the double-glazed panel on the frame with the exposed edge of overlapped film facing down. Press the panel firmly into the caulk, then drill ¼-inch holes at 12-inch intervals through the panel frame, counterboring for the heads of 2½-inch wood screws. Screw the frame to the wood below. Apply a bead of caulk around the edges of the panel and lay the second panel, glazed side up, atop the first panel, anchoring it with screws as before. Then caulk all exposed seams and screwheads.

**3** **Flashing the glazed panels.** Cut a strip of aluminum flashing 6 inches wide and the same length as the lower end of the tank frame; bend so that 2 inches of aluminum will extend across the top surface of the glazed panel and 4 inches will extend down the side. Anchor both aluminum surfaces to the wood frame with ¾-inch aluminum screws at 1-foot intervals. Cut flashing for both sides of the frame and attach it in the same way; then fashion a strip for the top. Use silicone caulk to cover all edges where the aluminum overlaps the acrylic-polyester film.

To fill the system, close the boiler drain valve at the bottom of the tank frame; open the gate valves on the hot-water and cold-water lines inside the house, and close the gate valve on the vertical line between them.

# A Solar Swimming-pool Heater

One of the most logical uses of solar energy is to heat a home swimming pool, since sunshine and swimming weather are so closely linked. The simplest and least expensive way to warm an unheated pool with sunrays, or to ease the load on a gas or electric pool heater, is to cover the water with a transparent plastic blanket, commonly made of heavy, double-thickness polyethylene. By preventing evaporation (which can account for up to 55 per cent of a pool's heat loss) while still permitting sunlight to warm the water, a plastic cover can raise the temperature of a pool by as much as 10° F. on a sunny day. Used in conjunction with a conventional heater, a pool blanket can pay for itself over a single season in reduced energy bills.

But full-fledged solar systems, available from solar equipment-dealers and from some pool-supply stores, are even more effective. Most pools are heated to around 80° F.; in warm weather, a solar system can do the job unaided by a standard gas or electric heater. In cooler weather it can save considerable energy by preheating the water before it passes into the conventional heater.

Because the basic plumbing necessary to circulate the water through the solar collector panels is already included in the pool's filter system, a solar heating loop need only tap into the existing pipes—usually at a point just beyond the filter but before the conventional heater, if any. A one-way check valve prevents water from draining back into the filter from the elevated collectors, and a manual or thermostatically-controlled gate valve allows pool water to bypass the solar heater on cloudy days, or whenever the pool has heated up sufficiently. The existing swimming-pool pump will often be powerful enough to circulate water through the panels, but the panel manufacturer or dealer may recommend that you retrofit a larger pump.

Solar collector panels for pool heating are simpler and lighter than the ones that are designed for hot-water systems, requiring no insulation or glazing. Many are copper; others are merely 3-by-8-foot sheets of molded plastic, honeycombed with channels through which the water circulates.

The panels must be mounted on a roof and oriented toward the sun in the same manner as other solar collectors. If nearby south-facing roof surfaces do not provide the proper angle (page 78), you can secure the panels to an inclined roof rack, similar to the one on page 80, but covered with ¾-inch plywood (the lightweight pool panels will sag in the middle if they are not supported). In many cases, however, the simplest way to mount the collectors is to build a nearby cabana-style shelter with its roof oriented due south and pitched at an angle equaling your latitude.

To heat the water adequately, you will need collector panels with a total surface area equal to half the area of your pool. To ease installation, use schedule-40 PVC pipe and, whenever possible, PVC valves and fittings to make the plumbing connections. Buy the type of pipe that is specifically labeled for outdoor use; the plastic is formulated with a stabilizer that enables it to withstand prolonged exposure to sunlight. The PVC pipe ordinarily used for indoor plumbing degrades rapidly in the sun and must be painted if it is used outside the house. You can buy pipe, fittings, and any other hardware not included with the panels, at plumbing and home-improvement stores.

## Installing Panels and Pipes

**1** **Mounting the collector panels.** Mark the position of each panel and the locations of the rafters on the roof (page 81, Step 1). Along the top and bottom guidelines, drill a ¼-inch hole into each rafter. Fill the holes with butyl rubber caulking. Haul the first panel to the roof (page 82) and position it between the guidelines. Anchor the bottom header pipe to each rafter by hooking a 1-inch one-hole pipe strap over the pipe and securing it with a 2-inch galvanized lag bolt, driven into the filled hole. Secure the top header pipe in the same way.

Mount the subsequent panels, connecting the abutting headers with 3-inch lengths of 1½-inch-diameter rubber hose slipped over the ends of the pipes and secured with hose clamps (inset). At each end of the row of panels, spread a ½-inch band of PVC solvent cement around the end of the pipe that will not be used as a water inlet or outlet, and twist on a plastic pipe cap.

Secure the midsection of each panel with three flexible nylon or rubber retaining straps, stretched horizontally across the panel, 2 feet apart. Fasten the straps with 2-inch lag bolts, screwed into caulked holes in the rafters.

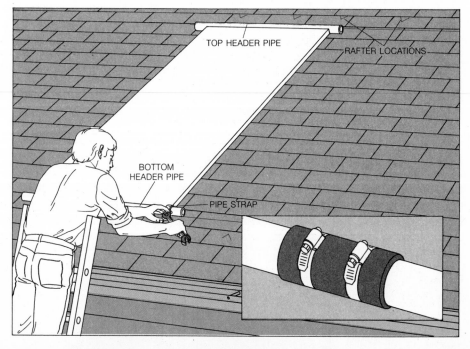

TOP HEADER PIPE

RAFTER LOCATIONS

BOTTOM HEADER PIPE

PIPE STRAP

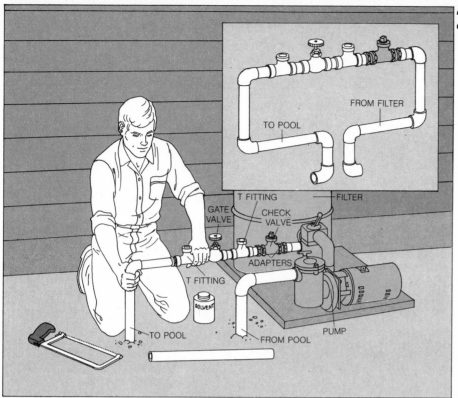

**2** **Rerouting the filter system.** Before tapping into the existing system, assemble the valves and fittings that will divert water to the collectors. Attach a 2-inch length of 1½-inch-diameter PVC pipe to each end of a PVC gate valve with PVC pipe cement. (If you are installing a thermostatically-controlled automatic valve, use plastic-to-steel adapters between the valve and the adjoining PVC fittings.) Bond a 1½-inch PVC T fitting to each end of the assembly. Then connect a check valve to the end of one T with a plastic-to-steel adapter. On the opposite side of the check valve, install a second plastic-to-steel adapter if the existing pool pipe is plastic. If it is copper, attach a copper-to-steel adapter instead, and add a plastic-to-copper adapter at the opposite end of the assembly.

If the run of pipe carrying pool water from filter to conventional heater or back to the pool is long enough, splice the valve-and-fitting assembly into the existing line, using reducers, if necessary, to match pipe diameters. Shut down the filter system, then use a hacksaw or tubing cutter to remove the necessary length of pipe, and bond the assembly in place with solvent cement. If you cannot fit the assembly into the existing pipe run, use elbows and pipe to construct a loop that is wide enough to accommodate the new valves and fittings *(inset)*.

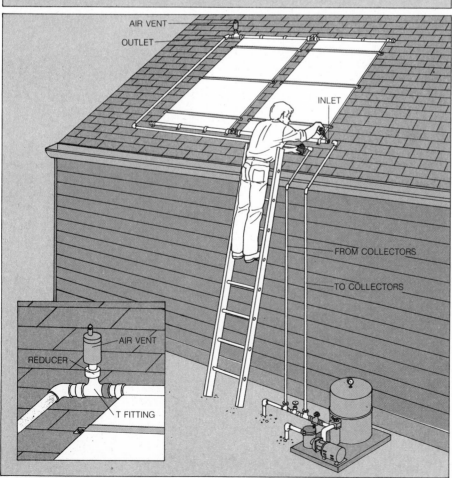

**3** **Linking the panels to the filter circuit.** Run 1½-inch PVC pipe from the T fitting farthest from the pool filter to the water outlet of the collector panels. Secure vertical runs longer than 8 feet, using plastic pipe supports spaced every 6 feet. Connect the pipe to the panel outlet with a T fitting. Attach a 2-inch length of pipe to the open leg of the T, then add a reducer of the appropriate size and an air vent *(inset)*. Complete the loop by running pipe from the second T on the filter line to the panel inlet.

To activate the system, close the gate valve; with an automatic valve, connect the temperature sensor leads to the points on the circuit recommended by the valve manufacturer. Turn on the pump to send water up to the collectors.

# 3

# A Diversity of Energy Sources

**A fast and efficient heater.** No bigger than a desk telephone, this energy-saving device heats water as it flows instead of in a large, heat-losing storage tank. Known as an on-demand heater, the device uses sophisticated electronic controls and a high-efficiency heating element, housed in the cylindrical chamber at left, to raise water temperatures as much as 90° F. Added to a water line near a sink, shower or dishwasher, the heater is then wired to an electrical circuit—a connection that can be made with a screwdriver *(page 110)*.

"In this world, nothing is certain but death and taxes," quipped Ben Franklin nearly two centuries ago. Now—particularly around the first of any winter month—homeowners are prone to add high utility bills to Franklin's twain. But while the Grim Reaper and the revenuer will always have their due, several methods of reducing—and in some circumstances eliminating—budget-busting utility bills are available to homeowners today.

Many of the devices gaining popularity, including wood-burning furnaces and heat pumps, warm the air of a house. Others, such as tankless water heaters and penny-pinching timers that carefully ration power to a conventional water heater, save money yet assure warm showers and clean clothes. Still others—photovoltaic cells, wind machines, and water turbines—produce free electricity, although they are costly to install initially.

The devices that heat air or water are frequently used in conjunction with solar collector systems, either increasing the effectiveness of the collectors or serving as backups for long sunless periods. A heat pump, for example, can be wired to turn on only when the solar heating system fails to keep a house at a comfortable temperature. Similarly, an on-demand water heater, which heats water as it is used rather than by the tankful, can give a final warming to water for, say, automatic dishwashing should the solar system leave the water too tepid to dissolve away grease.

Heating devices can also be interconnected with conventional utility-powered systems, not so much to back up a cheaper heat source but to be backed up themselves by more expensive utility power. Wood-burning and coal-burning furnaces, for example, provide reliable, economical heat; but they require regular attention. As long as the furnace is kept stoked, there is no necessity for utility power. If the fire dies, however, then gas, oil or electric heat must be ready to take over.

Linking unconventional and conventional energy sources is a possibility for home electricity as well. For those with resources—financial and natural—a windmill, water turbine or solar electric panels can produce electricity to handle domestic needs. On days when the wind dies or the stream dries up, an interconnection with the public utility or a bank of storage batteries is needed to take over. An interconnected home generating system pays off in another way as well: By United States law, a public utility must buy any power that a home generator feeds into the company's power lines. In a lucky month—when a homeowner's production exceeds the household's demand—the power company may have to pay a bill rather than send one on the first of the month.

# Wood and Coal: Veteran Fuels on the Comeback

For a time wood and coal nearly disappeared from the home, replaced by cleaner, more convenient heating fuels—oil, natural gas and electricity. Now wood and coal are making a comeback because they have advantages that are hard to ignore: the coal supply is plentiful, the wood supply renewable. What is more, heating with wood and coal is no longer limited to fireplaces and potbellied smoke-belchers.

New designs of stoves and furnaces—some burn only coal, some only wood, some both—give them much greater heat-transfer efficiency than older models. The new designs, which rely heavily on the precise regulation of oxygen sustaining the fire, allow owners to heat houses with less fuel and greater control.

Before putting old fuels and new technology to work, however, a potential owner must make careful plans and seek professional advice. A simple central-heating system, such as the one shown at right, is suitable for a small house or even a somewhat larger house if it can be supplemented by space heaters. It is fairly easy to plan. But a more complex system—one that must supply all the heat for a large house or one that will be teamed with an existing furnace—must be carefully selected, taking into consideration the heat loss of the house, the projected use of the new furnace, the capacity of any existing heating plant and the size of ductwork.

Before buying or installing a wood or coal heating system, check with local officials, especially fire marshals, concerning your local codes. The codes will specify clearances needed between heated parts of the system and combustible parts of a house. The clearances that the codes specify, however, are merely minimums; manufacturers may call for more distance. Always use the greatest clearance anyone recommends.

To heat with wood you will, of course, need a ready supply of firewood and a place to store it. Many wood-burning furnaces can accommodate logs up to 4 feet long. You will also need to dispose of the ashes—at first in a metal container until they are cool, then permanently. Fortunately, wood ashes make rich tree and garden fertilizer. An additional wood by-product requiring attention is creosote, a sticky residue that will accumulate in the flue. To prevent chimney fires, you must keep the creosote deposits to a minimum, generally less than ⅛ inch thick.

Although some manufacturers make wood- and coal-fired boilers for hot-water systems, the central-heating systems shown on the following pages are designed to circulate warm air. The first system (below) operates solely on convection, relying on the fact that heated air will rise. The heating appliance for such a system could be either a full-sized furnace or a compact stove with a sheet-metal jacket that traps the heat radiated by the firebox; both channel heated air into a piece of ductwork called a plenum, which attaches to other ducts leading to the room above.

The second system (page 101) is more elaborate. It has sheet-metal ducts to distribute the heat and a blower to move the warmed air. In this type of system the wood- or coal-burning furnace can stand alone and use a blower housed in its frame, or it can be spliced into the ducts of an existing furnace and work in tandem with that heating plant. In the latter case, the new appliance can sometimes use the blower of the original furnace.

Whichever system you select, be sure all components are labeled as having been tested by a nationally recognized laboratory for the usage you intend. Also be sure to obtain detailed installation and operation instructions.

To install either system, you will have to work with sheet metal and ducting. If you do not feel qualified to fabricate the sheet-metal parts you need, give exact specifications to a sheet-metal contractor. Detail for the contractor the proper gauges of metal and sizes of ducts demanded by the codes and manufacturers. Also discuss how each of the joints in the ductwork will be made. Some joints will require sheet-metal screws; others need special fasteners called S-slips and drive cleats (opposite). For these joints the contractor will have to bend flanges on the appropriate sections of ducting.

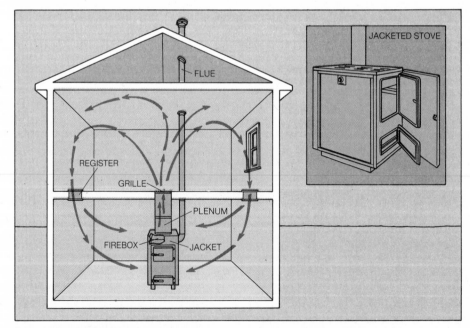

**A convection heating system.** A wood or coal furnace or a jacketed stove (inset) can heat a small house through one centrally located floor grille, as shown in this schematic illustration. Air, passing between the firebox and the sheet-metal jacket of the furnace, is heated and rises into the plenum, atop the jacket. From there, the air rises through ductwork—often specified by codes to be of a certain length so that the grille will not become too hot—and exits through the floor. The warmed air spreads and cooler air returns to the basement through registers in the floor. A flue carries smoke from the firebox out of the house.

# Floor Vents for a Wood-burning Furnace

**1 Framing an opening for the grille.** Working to dimensions specified by the grille manufacturer, and being careful to meet all clearance requirements, cut a hole through the floor. Align one side of the opening with the edge of a joist. If the opening must pass through a joist, saw the joist back 3 inches from the edges of the cuts. You will need to do much of this cutting with a handsaw. To support the ends of the cut joist, nail 3-inch-wide joist hangers to the joists on each side and install two header joists one at a time in the hangers. Nail through the first header joist into the end of the cut joist; nail through the second header joist into the side of the first.

**2 Installing the ductwork.** After positioning the furnace, slip the plenum over the top hole in the furnace jacket and fasten it in place with sheet-metal screws. Next, screw the grille to the top of an extension box—a piece of ductwork that penetrates the floor—and lower the assembly into the floor opening until the flange of the grille rests on the floor surface.

In the basement, check the alignment of the remaining pieces of ductwork.

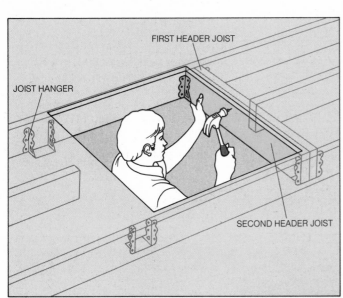

FIRST HEADER JOIST

JOIST HANGER

SECOND HEADER JOIST

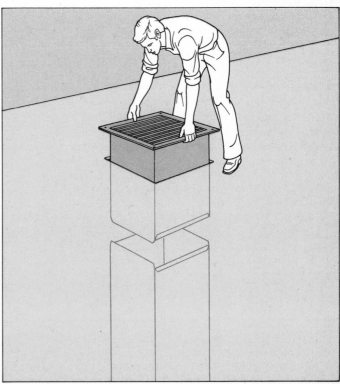

**3 Joining the ducts.** To join two pieces of rectangular duct, slide S-slips onto the straight edges of one *(near right)*. Then slide the straight edges of the second duct into the open sides of the S-slips. Next, link the two duct pieces by sliding drive cleats over the duct edges that have been bent back *(far right)*. Fold the projecting ends of the drive cleats around the corners of the duct to overlap the ends of the S-slips.

To complete the heating system, cut holes for cold-air return registers through the floor near outer walls of the house. Insert registers in the holes and screw them to the floor. If the basement has a finished ceiling, install registers in the ceiling as well. Use short extension boxes to join the ceiling registers to those in the floor above.

DRIVE CLEAT

S SLIP

S SLIP

DRIVE CLEAT

# Adding On a Forced-air Furnace

A wood- or coal-burning furnace can be grafted onto the ducts of an existing forced-air system in one of three different configurations. For the sake of brevity, only the addition of a wood-burning unit will be described on these and the following pages. But the procedures for adding on a coal-burning furnace would be identical.

In a downstream installation, cool air returning from the rooms of a house (return air) passes through the old furnace before entering the wood furnace. In a parallel installation, return air is diverted from its course to the existing furnace and routed to the wood burner. An upstream installation has return air passing through the wood furnace before it reaches the old heating plant.

The upstream installation is not recommended since air heated by the wood fire can cause the blower motor of the old furnace to overheat. The parallel installation also has a drawback: It requires an additional blower for the wood furnace and, thus, an additional expense. But the shape and area around the existing heating plant may prevent a successful downstream installation and the parallel configuration will be your only choice. Also, some downstream installations need an additional blower themselves.

Enlist the aid of a professional forced-air heating engineer in planning the system for your house. Unfortunately, an add-on furnace increases air resistance through the entire system. This puts an extra burden on the blower of the existing furnace and changes the rate at which air moves past the fireboxes to be warmed. Only a qualified technician can properly assess your existing system to determine if its ducts and blower are adequate for an add-on wood furnace. The heating engineer can also tell you what complications your air-conditioning equipment may pose to an expanded heating system.

Determining the size of the add-on furnace is also an important step. Even though on some days the wood burner may heat the house by itself, it does not require as high a BTU rating as the old furnace does. For one thing, the wood-burning furnace will always have a dependable backup; for another, most oil and gas furnaces have greater heating capacity than is actually necessary, as a measure of insurance against severe cold spells. Ideally, the wood-burning furnace should have a BTU rating slightly less than the BTU loss from the house on the coldest days of the year—a larger furnace would run at less than capacity on most days and permit too much creosote build-up in the flue.

In selecting a wood burner to suit your needs, be wary of salesmen's claims. The salesman's estimation of furnace capacity will probably be based on the manufacturers' BTU figures, which are calculated under ideal laboratory conditions. To make your own estimation of a wood burner's capacity, use the formulas in the box at right.

An expanded heating system can incorporate varying degrees of automation. A simple downstream system can function with a single, manually operated duct damper and a coiled bimetal thermostatic strip. The damper diverts return air to the wood-burning furnace; the bimetal strip reacts to heat above the wood fire and raises or lowers a smaller damper that controls combustion.

A more complicated installation would include a thermostat in the heated portion of the house, an electric motor in place of the bimetal strip and a sensor that turns on the blower when the air above the firebox reaches a predetermined temperature. A parallel installation also requires specially balanced duct dampers that open and close automatically, depending on which heating appliance is operating at a given moment.

## Estimating the Usable Output of a Wood Furnace

Because it is important to match the size of a wood furnace to the needs of your house, it is advisable to take a hard look at manufacturers' claims of BTU output. To calculate how many BTUs a furnace can deliver per hour, begin by measuring the usable volume of the firebox. In doing so, include only the parts of the firebox where logs can reasonably fit. You cannot, for example, usually place logs above the height of the firebox door.

Multiply the volume of the firebox by the density of the firewood. If you plan to use medium-sized pieces and pack them in with moderate tightness, use a factor of 15 pounds per cubic foot for softwood, 18 for hardwood. If you use small pieces and pack them very tightly, use a factor as high as 22. Next, multiply the result by 6,800, the average BTU content of 1 pound of seasoned wood. Multiply that figure by an efficiency factor of .55. This will give you a good estimate of the BTUs one charge of wood will deliver to your house. Divide the result by a reasonable estimate of how long a full charge of wood will burn—usually six to eight hours. This final figure represents the usable BTU output of the furnace during one hour of operation.

By comparing that calculation to an estimate of the number of BTUs it takes to heat your house for one hour, you can determine whether a furnace you are considering will be the optimum size for your house.

**Two expanded heating systems.** The forced-air heating system shown below, left, has a wood-burning furnace that has been added on downstream from the original furnace. A manually operated damper in the plenum of the old furnace controls the flow of air. When a wood fire is burning, the damper is closed. Thus, return air is forced through the original furnace—which is not running—and is channeled into the wood furnace. There the air is heated and blown into the supply ducts for the house. In this example, the blower in the original furnace is adequate to move air through the entire system. The blower runs continuously during the time that a wood fire burns, but can be switched to an automatic setting that turns it on and off while the old furnace is operating.

The heating system shown below, right, is a parallel configuration in which both furnaces have their own blower units. Return air can thus be channeled through either furnace without passing through the other one. Arrows indicate the path of the air while a wood fire is burning. When the wood fire goes out or cannot maintain adequate warmth in the house, a thermostat turns on the original furnace. Two carefully counter-balanced dampers control the flow of air through the system. The dampers automatically open or shut, depending on which blower unit is running.

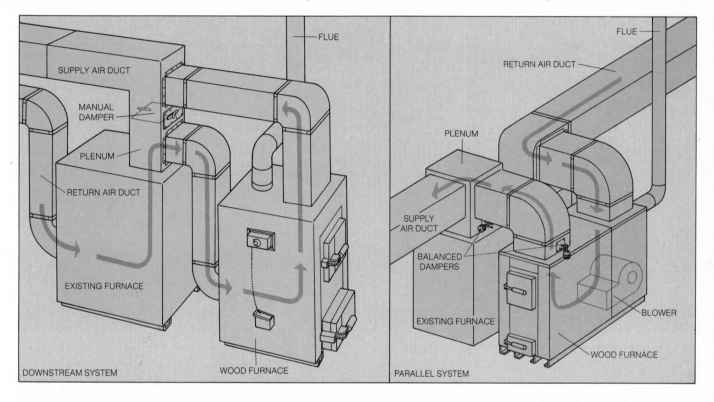

DOWNSTREAM SYSTEM

FLUE
SUPPLY AIR DUCT
MANUAL DAMPER
PLENUM
RETURN AIR DUCT
EXISTING FURNACE
WOOD FURNACE

PARALLEL SYSTEM

FLUE
RETURN AIR DUCT
PLENUM
SUPPLY AIR DUCT
BALANCED DAMPERS
EXISTING FURNACE
BLOWER
WOOD FURNACE

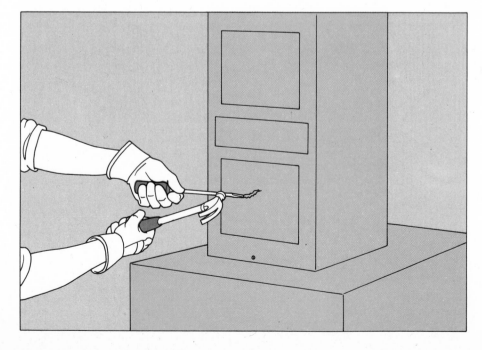

## Linking Furnaces with Dampers and Ducts

**1 Cutting into the plenum.** Mark the plenum of the existing furnace for three rectangular holes—two for the ducting that will tie in the new wood burner and the third for the manual damper that you have acquired for your new system. Make the duct holes the same dimensions as the connecting pieces, called take-offs; make the damper hole just large enough to permit the damper to pass through. Pierce the metal inside a marked outline by holding the tip of an old screwdriver or a cold chisel against the sheet metal and striking the shank with a hammer. Make a hole large enough to admit the tips of a pair of tin snips, then cut along the marked lines. Cut out the other openings in the same manner.

**2** **Installing the damper.** Either buy or make a damper *(inset)* that will fit snugly inside the plenum. The shanks on which the damper will swivel have pronged clamps at one end that fasten them to the edges of the damper plate.

On the side of the plenum opposite the damper opening, drill a hole just larger than the damper shanks. Pass the damper into the plenum through its opening, then insert one of the shanks in the drilled hole and fasten it with a wing nut. Cut a rectangular piece of sheet metal slightly larger than the damper opening and, at its center, drill a hole for the other shank of the damper. After fitting that shank into the hole, use sheet-metal screws to fasten the new plate to the plenum walls, covering the damper opening. Slide the damper handle onto the square of the shank and tighten a wing nut on that side.

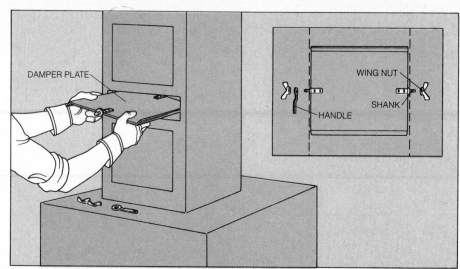

DAMPER PLATE

WING NUT

HANDLE

SHANK

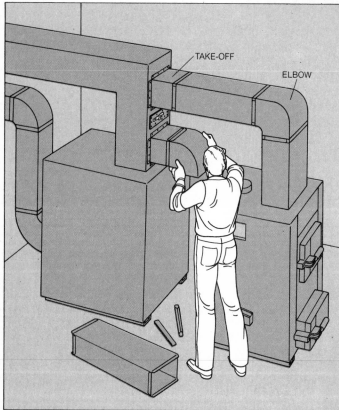

TAKE-OFF

ELBOW

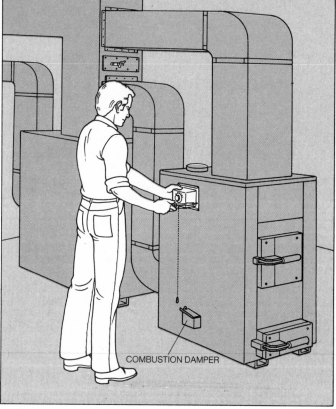

COMBUSTION DAMPER

**3** **Adding ducts.** Connect the add-on furnace to the existing furnace with new ducts consisting of take-offs, elbows and straight sections of sheet-metal ducting. Screw the take-offs in place over the openings cut in Step 1 and over the holes manufactured in the jacket of the wood-burning furnace. Add elbows to the take-offs and straight duct sections between the elbows, joining the pieces with S-slips and drive cleats.

When the new ducts are in place, have a professional heating engineer test the system to determine whether the rate of air flow meets the specifications of both furnaces.

**4** **Attaching the furnace controls.** Screw the control box for the combustion damper to the body of the furnace through the opening in the jacket directly above the damper. Attach the chain that comes with the control box to the damper door. A bimetal strip inside the control box expands and contracts with heat, thus raising and lowering the chain and controlling the flow of air into the fire. Install a flue as described opposite.

# Building the Flue

**1** **Installing the chimney-pipe support.** At the first floor through which the flue must pass, install a chimney-pipe support to bear the weight of insulated chimney pipes. To determine where to position the support, mark the basement ceiling—between the joists—at a point either directly over the center of the flue cutout atop the furnace or offset to allow for bends. Cut through the ceiling and floor. Then, working from the floor above, build a frame with lumber the same size as the joists. Nail two frame pieces perpendicular to the joists, and nail two flat against the joists. From the basement, raise the chimney-pipe support into the opening and check it for fit—the cylinder of the support should touch the frame on all four sides. With a helper holding the bottom of the support against the basement ceiling, check the top with a level and then nail through the cylinder into each of the four frame pieces.

CHIMNEY-PIPE SUPPORT
STOVEPIPE

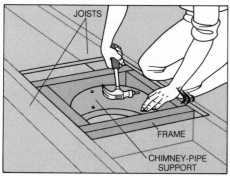
JOISTS
FRAME
CHIMNEY-PIPE SUPPORT

**2** **Running the pipe.** Join sections of single-wall stovepipe to make a piece long enough to extend from the furnace to the inside of the support at the ceiling; add elbow sections, if necessary. Fasten each stovepipe joint with three sheet-metal screws. Insert the top into the chimney-pipe support and fit the bottom to the cutout on the furnace. Attach the chimney-pipe support to the top of the pipe with four sheet-metal screws; in similar fashion fasten the bottom end to the furnace cutout.

At each floor above the first, use a plumb bob to mark centers for the openings through which the flue will pass. Cut the openings and install the metal spacers supplied by the flue manufacturer. The spacers ensure the proper clearance between the flue and the wood framing of the house. To mark the location for a hole through the roof, hang a plumb bob over each of the four corners of the opening below. Drill a small hole through the roof at the four corner points. Then, from atop the roof, cut out the shingles and sheathing between the marker holes.

FLASHING OUTLINE
FLASHING
STORM COLLAR
CHIMNEY CAP

**3** **Completing the flue.** Center the flashing—the metal fixture that supports the flue above the roof—over the roof opening and trace its outline on the shingles. Remove the flashing and draw a second outline 3 inches inside the first, across the top and two thirds of the way down the sides (*above, left*). Cut through the shingles touched by the inner outline and remove the cut pieces around the roof opening. Spread roofing cement on the shingles in the lower third of the area bounded by the large outline, then

move the flashing back into place, sliding it up under the edges of the cut shingles. Press the lower edge of the flashing into the roofing cement and nail through the flashing into the roof with galvanized nails at 3-inch intervals. Cover the nail-heads with roofing cement.

Assemble sections of chimney pipe to build up the flue from the chimney-pipe support to the opening in the roof. Raise the top of the flue through the flashing. If it is required by your

local building codes, enclose the chimney pipe with framing and wallboard.

Slip a storm collar over the chimney pipe so that it rests on the flashing (*above, right*). Seal the joint between the collar and the pipe with roofing cement. Add sections of chimney pipe until the pipe rises 3 feet above the high side of the roof where it exits, or 2 feet above any surface within a horizontal distance of 10 feet. Add a chimney cap to finish the installation (*inset*).

# Electric Controls for a Wood Furnace

Electric controls can automate a two-furnace heating system and make it much more convenient to use. However, the degree to which automation is possible will vary from system to system. In the heating system shown below, for example, a wood-burning furnace has been added downstream from the original oil burner; the system functions with a single blower, often called "fan" on controls. This equipment can be improved by the addition of a thermostat that will make the wood burner responsive to temperatures in the living quarters. But even with this addition you will still need to manually close the duct damper to the wood furnace when the fire dies.

A system with two blowers, on the other hand, can be fully automated—except, of course, for adding firewood. The two-blower system will have counterbalanced dampers that are opened and closed by the force of the blowers and thus save trips to the basement.

The controls for the heating system will have similar components, no matter what the configuration of the furnaces. Fan switches, linked to bimetal sensors in plenums, turn the blowers on and off at predetermined high and low temperatures. Limit switches, usually connected to the same sensors as the fan switches, are safety devices used to prevent abnormally high plenum-air temperatures. If the temperature rises past a preset point, the limit switch cuts off power to the gas- or oil-furnace burner.

The thermostat for an add-on wood burner will control a motor that adjusts the combustion damper. But the damper is also adjusted by a heat-sensing switch, similar to the fan and limit switches, that keeps the fire burning at a steady rate. The thermostat is therefore wired through the heat-sensing switch. Also, since the original furnace in your heating system will have a thermostat of its own, the new thermostat will have to be set 4° higher than the old one to keep the old furnace inactive while a wood fire is producing sufficient heat.

The wiring for these controls involves both low- and high-voltage circuitry. Thus the electrical system must incorporate transformers that reduce the incoming 120 volts to 24 volts to supply the low-voltage circuits. You should be certain your wood-burner combustion damper will close automatically in the event of a power failure.

All of the wiring must be done in accordance with local codes and is subject to inspection once the job is completed. The wiring techniques depicted here will be applicable to most add-on wood furnaces. But the particular heating system shown here is merely an example, and the configuration of the various electrical components may be different from the ones needed for your system. Follow the manufacturer's wiring diagrams to install controls on your own equipment.

**1** **Installing the hardware.** Slip the rods that contain the bimetal sensors for the damper control, the fan switch and the limit switch into the holes already provided in the jacket of the wood-burning furnace. Attach the body of the control assembly with screws. Then run the wires that are connected to the control assembly to the damper motor by attaching the mounting brackets on the wires to the jacket of the furnace.

Run low-voltage wires to the new thermostat location, adjacent to the thermostat of the existing furnace. Mount the base of the new thermostat on the wall, over the ends of the low-voltage wires. Use a small level to make certain that the thermostat base is absolutely level before tightening the mounting screws.

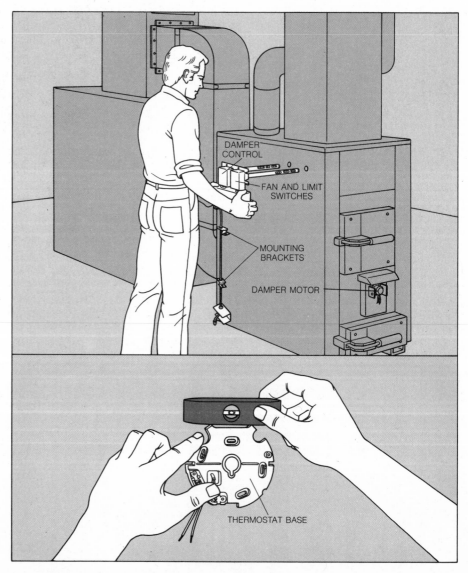

DAMPER CONTROL

FAN AND LIMIT SWITCHES

MOUNTING BRACKETS

DAMPER MOTOR

THERMOSTAT BASE

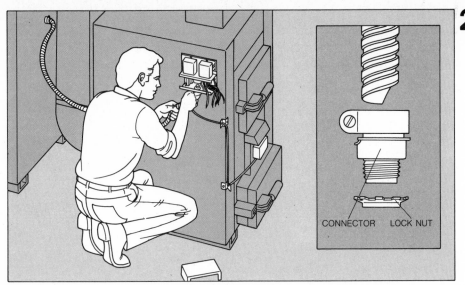

**2** **Adding flexible conduit.** Cut a length of flexible conduit—sometimes called Greenfield— to extend between the control assembly on the wood furnace and the junction box of the original furnace. Connect the conduit to the two boxes with connectors and lock nuts (*inset*).

CONNECTOR    LOCK NUT

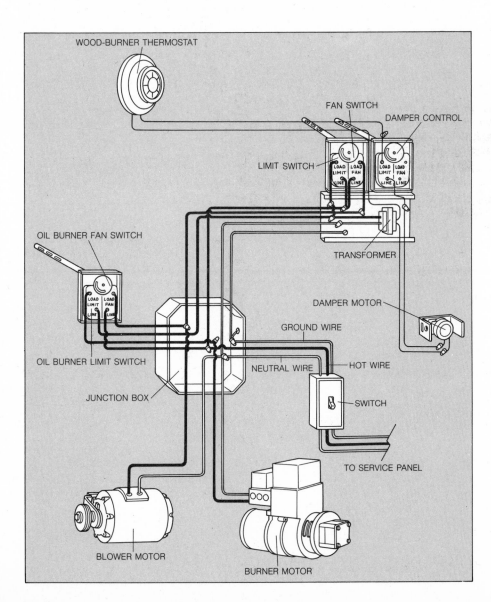

WOOD-BURNER THERMOSTAT

FAN SWITCH

DAMPER CONTROL

LIMIT SWITCH

LOAD LIMIT LINE / FAN LINE

LOAD LIMIT LINE / FAN LINE

TRANSFORMER

OIL BURNER FAN SWITCH

LOAD LIMIT LINE / FAN LINE

DAMPER MOTOR

GROUND WIRE

OIL BURNER LIMIT SWITCH

NEUTRAL WIRE    HOT WIRE

JUNCTION BOX

SWITCH

TO SERVICE PANEL

BLOWER MOTOR

BURNER MOTOR

## Mapping Out the Circuits

**Wiring the controls.** After feeding No. 14 or 12 wire through the conduit installed in Step 2, use wire caps to make the connections shown in this schematic diagram. Wires of the original furnace that are not affected by new circuitry are not shown. The objects of the new wiring are threefold: to activate the motor on the combustion damper; to allow the oil-burner motor to operate only when both limit switches are closed; and to run the blower motor when either of its fan switches closes.

# Heat Pumps: Devices That Both Heat and Cool

The economical heat pump, which uses ambient heat to both heat and cool a house, is an increasingly popular alternative to more conventional heating and cooling systems—sometimes replacing them, sometimes serving as an auxiliary system. Similar in function to an air conditioner except that its role can be reversed, the heat pump in one mode transfers heat into the house, and in another mode extracts it. Furthermore, a single thermostat setting senses which mode is required and reverses the heat pump automatically.

The key ingredient in a heat pump is the refrigerant in its coils, usually a substance called Freon, which vaporizes into a gas at a boiling point far lower than the 212° F. that water requires. When the refrigerant boils, changing from a liquid to a gas, it absorbs heat from its surroundings; it can do so even when those surroundings are relatively cool. As the refrigerant changes back into a liquid from a gas, it gives up its heat to the atmosphere. This process of transformation from liquid to gas and back again is controlled by an expansion valve and an electric compressor.

Heat pumps are most efficient in the cooling mode. As a source of heat their efficiency varies according to the outdoor temperature. When the temperature dips below about 40°, heat pumps must draw more heavily on supplemental electrical resistance coils built into the unit. In cold climates, heat pumps can be effective as an alternative source of heat, installed in tandem with an existing fuel-fired heating system and hooked to a thermostat that switches from one system to the other whenever the outdoor temperature is high enough for the heat pump to function economically.

Heat pumps can be cheaper to operate than other heating systems because, by tapping into the free heat in outdoor air, they give back more energy—in the form of heat—than the equivalent amount of electrical energy they consume. Thus they are more efficient and—under optimal conditions—less expensive to operate than gas or oil furnaces or electrical

resistance systems; the amount of energy, expressed in BTUs, that is gobbled up by these conventional systems always exceeds the BTUs of heat they produce.

Heat pumps are steadily gaining popularity, but they are not for everybody, because not everybody has the optimal conditions that guarantee a good payback (page 8) for the initial cost of the pump. The long-term saving that a heat pump can offer will depend on climate, house construction, insulation, and the relative cost of other sources of heat in a particular area.

In rating heat pumps for both heating and cooling capacity, manufacturers use two terms that the potential purchaser should be familiar with. Both compare the ratio of energy input to energy output. For the heating mode, the ratio is called the coefficient of performance, or COP, calculated by dividing BTU input into BTU output; the COP of a typical heat pump can range from 2.8 at 60° outdoor temperature to 1 at 10°. For the cooling mode, the ratio is called the energy efficiency ratio, or EER, a figure that is determined by dividing the unit's wattage into the BTUs removed; typically, the EER can range from 6 to 10. The higher the EER and the COP ratios, the more efficient the unit.

Also to be considered is the source of heat energy tapped by the heat pump—air, water or even the sun. Those that use air, called air-to-air heat pumps, are available in one-, two- and three-piece systems. One-piece systems, called single-package units, are similar to room air conditioners. These units sit in a window or are installed outdoors, usually on the roof or at the side of the house, and are connected directly to the house ductwork. Two- and three-piece systems, called split systems, have an outdoor section connected to one or two indoor sections by pipes running through the house wall. The indoor section can be tied into an existing forced-air heating system to share the same ducts and blower, in much the same way as central air conditioning is connected.

Heat pumps that tap the energy of wa-

ter, called water-to-air systems, are connected to a well or to a pond that does not freeze over. Water-to-air heat pumps are one-piece systems, drawing water from the well or a pond and expelling the used water into a discharge well or into a lawn sprinkler system. Because their source of heat remains at a fairly constant temperature—about 50°—throughout the year, water-to-air systems are often 25 per cent more efficient than air-to-air systems, although they can be more expensive to install.

One variation of a heat pump, more modest in scale than the kind used for house heating and cooling, can heat the domestic water supply. Actually a cabinet-sized air-to-water system, this small heat pump absorbs heat from the indoor air and transfers it to the water tank. The unit maintains the water at a temperature of 120° to 140° while lowering the temperature of the surrounding air only 1° or 2°. In addition, this type of unit serves as a dehumidifier.

A heat-pump water heater monitors water temperature in the tank by means of a sensor, and the unit is automatically activated whenever hot water is withdrawn and replaced by cold water. In most cases, no backup heat source is needed, but the existing heat source—oil, gas or electricity—remains in place and can be turned on if the heat pump is closed down for servicing.

A typical heat-pump water heater is sold as a kit containing the heat-pump cabinet, fittings for the inlet and outlet shutoff valves and, in some models, a new pressure-relief valve with a longer sensing tube to replace the tank's existing short-tube pressure-relief valve. Hoses for the pump's water connections and for the drain line that carries off condensation are included, but in some localities these flexible hoses will have to be replaced with rigid piping; check your local plumbing codes. The unit plugs into a 20-ampere, 115-volt circuit like that required for central air conditioners and heavy appliances. If there is no such circuit in the vicinity of the unit, one will have to be installed at the main electrical panel.

# The Magical Mechanics of a Heat Pump

**An air-to-air system.** In the heating mode, cold refrigerant liquid, vaporized to a mist under low pressure, passes through the outdoor coils, absorbs heat from the air and changes from a mist to a gas. The gas is pulled through the reversing valve into the compressor, which makes it denser and hotter. The hot gas is then pumped to the indoor coils, giving up its heat to the air circulating around the coils. The heated air, blown past supplementary heating coils out of

sight near the blower, is distributed through the house ducts. The refrigerant emerges from the indoor coils as a liquid. It passes through a low-pressure expansion valve, where it changes back into mist, drops in temperature, and enters the outdoor coils to repeat the cycle.

In the cooling mode, the direction of the flow of refrigerant is changed by the reversing valve: The refrigerant in mist form first flows through the

indoor coils to pick up heat from the air. Following a series of changes exactly opposite those of the heating mode, the refrigerant then expels heat outside, aided by an exhaust fan.

In three-piece systems, the compressor is located indoors in a separate cabinet and is connected by refrigerant pipes to the outdoor and the indoor coils. In one- and two-piece systems, the compressor is located outdoors.

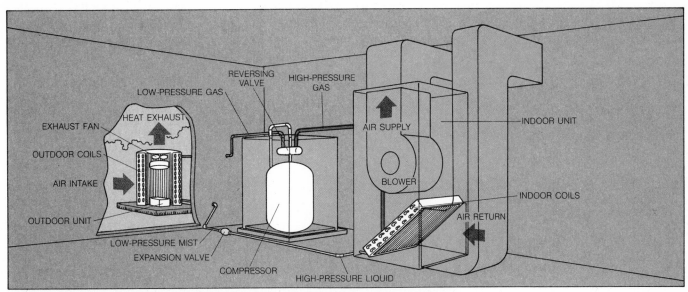

EXHAUST FAN
OUTDOOR COILS
AIR INTAKE
OUTDOOR UNIT
HEAT EXHAUST
LOW-PRESSURE GAS
REVERSING VALVE
HIGH-PRESSURE GAS
AIR SUPPLY
BLOWER
INDOOR UNIT
INDOOR COILS
AIR RETURN
LOW-PRESSURE MIST
EXPANSION VALVE
COMPRESSOR
HIGH-PRESSURE LIQUID

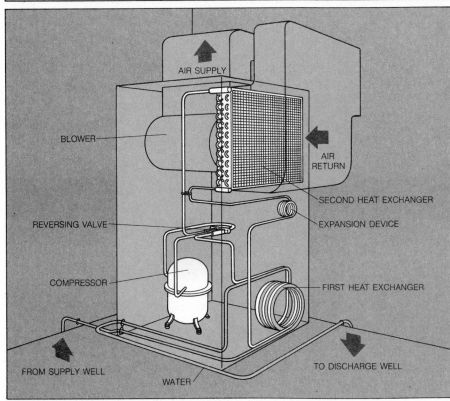

AIR SUPPLY
BLOWER
REVERSING VALVE
COMPRESSOR
FROM SUPPLY WELL
WATER
AIR RETURN
SECOND HEAT EXCHANGER
EXPANSION DEVICE
FIRST HEAT EXCHANGER
TO DISCHARGE WELL

**A water-to-air system.** In the heating mode, water from the supply well is pumped into pipes that circle past the refrigerant coils of the first of two heat exchangers, which are similar in function to the coils of an air-to-air system. The refrigerant in the coils of this first heat exchanger absorbs heat from the water and changes from a mist to a gas. The water, having given up its heat, is discharged into a well or a sprinkler system. The compressor makes the gas denser and hotter and then pumps it to the second heat exchanger, where the gas releases heat to air that will be distributed through the house. Emerging from the second heat exchanger as a liquid, the refrigerant is changed to a mist by an expansion device and returns to the first heat exchanger to repeat the cycle.

In the cooling mode, the path is reversed. After absorbing heat from the indoor air, the refrigerant passes heat to the water circling past the coils of the first heat exchanger's pipes. The water is then discharged.

# A Pump for Heating Water

**An air-to-water heat pump.** The components of this system are similar to those of the water-to-air heat pump (*page 107*). Cool water from the bottom of the tank is pumped into the tube-within-a-tube heat exchanger—an inner tube of water surrounded by an outer tube of refrigerant. There, the refrigerant, heated to a gas by the compressor, loses its heat to the water, condenses to a liquid, and passes through an expansion valve to become a mist that is lower in temperature than the surrounding air. As it absorbs heat from the air, the refrigerant passes through evaporator coils and returns to the condenser as a liquid. The cycle continues until all the water in the tank is heated to the temperature set by the thermostat. Meanwhile, water in the air condenses on the outside of the cool evaporator coils and drips into a pan, from which a tube leads it away to a drain.

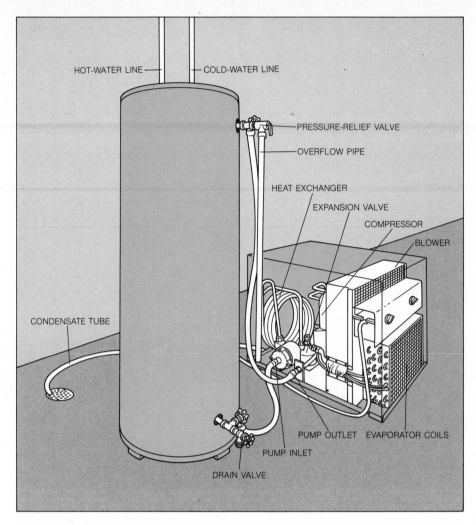

HOT-WATER LINE — — COLD-WATER LINE

PRESSURE-RELIEF VALVE

OVERFLOW PIPE

HEAT EXCHANGER

EXPANSION VALVE

COMPRESSOR

BLOWER

CONDENSATE TUBE

PUMP OUTLET    EVAPORATOR COILS

PUMP INLET

DRAIN VALVE

# Joining the Pump to the Heater

**1 Draining the water tank.** Shut off the heat source to the water tank: On an electric tank, turn off the power at the main electrical panel; on an oil-fired tank, turn off the burner switch; close the gas valve on the supply line of a gas-fired tank. Connect a garden hose to the drain valve at the bottom of the tank, and run the hose outdoors or to a floor-level drain or sump pump. Turn off the cold-water supply valve at the top of the tank and open all the hot-water faucets in the house. Open the drain valve, and drain the tank to the level of the drain valve. Disconnect the garden hose; remove the drain valve.

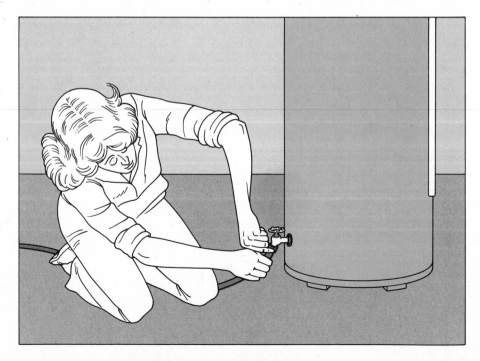

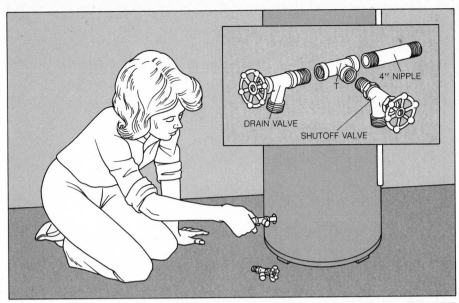

**2 Installing the tank inlet valve.** Thread a 4-inch nipple *(inset)* into the drain-valve opening, and screw the T fitting to the other end of the nipple. Screw the original drain valve onto the T end, and thread the shutoff valve on the center tap of the tee. To ensure watertight joints, first coat the threads with pipe-joint compound or plastic joint tape, and do not overtighten the threaded sections.

In inset labels: 4" NIPPLE, DRAIN VALVE, T, SHUTOFF VALVE

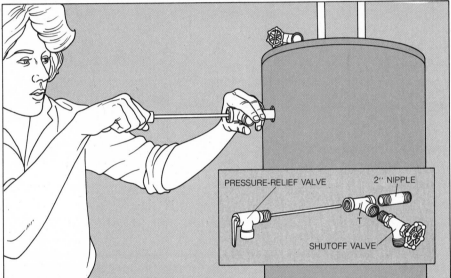

**3 Installing the tank outlet valve.** Unscrew the overflow pipe from the existing pressure-relief valve, located on the side or top of the tank, and set the pipe aside. Remove the pressure-relief valve, and replace it with the 2-inch nipple of the new outlet fitting *(inset)*. Thread the T fitting onto the nipple, then thread the new pressure-relief valve onto the T end. Thread the shutoff valve onto the center tap of the T fitting, and insert the overflow pipe in the new relief valve.

In inset labels: PRESSURE-RELIEF VALVE, 2" NIPPLE, T, SHUTOFF VALVE

**4 Connecting the hoses.** Remove the cap plugs from the pump inlet and outlet. Connect one hose to the pump outlet and to the shutoff valve at the top of the tank. Connect the other hose to the bottom shutoff valve on the tank and to the pump inlet.

Connect the clear plastic tubing to the condensation drain on the back of the pump, and run the tubing to an indoor floor drain, to a sump pump or to an outdoor location.

Turn the two shutoff valves to the fully closed position; then turn on the tank's cold-water supply valve. Allow the tank to fill until water flows in a steady stream from the open hot-water faucets throughout the house. Turn off the faucets and open the tank shutoff valves. Plug the pump into a 20-amp, 115-volt circuit and turn it on.

# Hot Water in a Hurry

Except for the furnace or boiler that heats the house, the largest energy consumer in most homes is the water heater. Typically it keeps a 40- to 60-gallon tank of water hot 24 hours a day. Yet the average household uses hot water no more than two hours a day. Worse still, in homes with dishwashers, the temperature must be held 30° to 40° F. above the normal 100° temperature used for showers and sinks; otherwise, cooking fats and greases will not dissolve.

Europeans long ago remedied this energy-wasteful setup with on-demand water heaters, compact devices made without the traditional storage tank. On-demand heaters use highly efficient heating coils and heat-exchange chambers to raise water temperatures only when the water is turned on. Now available in the United States and Canada, these heaters can be fueled by natural or propane gas or by electricity. The larger gas-fired models can meet the hot-water needs of a small house or a vacation cottage, but they are expensive. Also, the natural-gas versions must be connected to gas lines by a professional plumber.

The smaller electric models (right, top) generate only enough hot water to service a single plumbing fixture or appliance, but they are cheaper and can be installed by anyone familiar with basic plumbing and wiring techniques. However, you may need an electrician's guidance in connecting the power circuit to the main electrical panel.

Small electric on-demand heaters can be a convenient solution if you need additional hot water for a new bathroom or laundry room. They can also be used in conjunction with an existing water heater to boost water temperatures at specific locations. For instance, if you install one near a dishwasher, you can turn down the temperature at the main water heater to 100° and raise it to 140° just as the water enters the dishwasher. Small on-demand heaters also reduce the heat lost by hot water traveling through long lengths of pipe.

Although on-demand heaters save energy costs, the exact amount they save depends on several variables, including the type of heater you choose, the type of heater you currently use and the relative cost of the various forms of energy in your area—gas, oil, electricity. Manufacturers estimate that you can save 22 per cent or more by switching from a conventional electric water heater to a series of on-demand electric heaters. If you use an on-demand heater to supplement an existing heater, you can probably cut your heating costs about 5 per cent for each 10° you lower the temperature at the main heater.

Savings in energy will also come from an inevitable change in water-use habits. On-demand heaters raise water temperatures in inverse ratio to the rate of water flow: The faster the water flows through the unit, the less heat it picks up. At 2 gallons a minute, the maximum rate of flow through most electric on-demand heaters, the water temperature will rise only 30°; at the slower rate of ¾ gallon a minute, the temperature will soar by 90°. In winter, when the water entering the unit is colder, the flow rate may have to be decreased even further to achieve suitably high temperatures.

To accommodate the lower flow rates, you will have to install low-flow shower heads, which can operate satisfactorily on less than 2 gallons of water a minute, and low-flow faucet aerators, which require as little as half a gallon a minute. By putting a squeeze on the flow, these devices save water as well as the energy used to heat it.

Since many on-demand heaters are manufactured overseas, make sure that the installation procedures for the unit you choose comply with local building codes in the United States or Canada. If in doubt, check with your local plumbing or electrical inspector. In addition, make sure that gas models have been certified by the American Gas Association or the Canadian Gas Association, and that electric models bear the Underwriters Laboratories or Canadian Standards Association seal of approval.

**How an on-demand heater works.** With its plastic cover removed, an on-demand water heater looks more like a small engine part than a sophisticated electrical device capable of heating 100 gallons of water per hour. The key to this remarkable efficiency is the unit's cylindrical heat-exchange chamber, shown with a portion of its wall cut away to reveal the coiled, copper-sheathed heating element inside. The heating element draws current from the terminal block, to which the electric power cable is connected. Water entering the unit through the water inlet automatically activates a flow switch to complete the electric circuit.

The water enters the heat-exchange chamber through a pipe at lower right and is heated by the copper coil as it rises to the top of the chamber. There, an open-ended copper drain tube draws off the hot water, channeling it down through the center of the heating element and through the hot-water outlet. The unit's performance is regulated by an electronic heat-sensing element, or thermistor, at the top of the chamber. The thermistor monitors the temperature of the water, comparing it with the temperature that has been set on the unit's thermostat adjustment dial. If the desired temperature can be reached without constant

consumption of power, the thermistor overrides the flow switch and shuts off the power, switching it back on only as necessary.

A thermal cutout, also mounted at the top of the heat-exchange chamber, prevents water in the chamber from overheating if the water supply is interrupted for any reason. The cutout shuts off all power if the temperature of the heating coil reaches 190° or the water temperature exceeds 160°. Additional protection against temperature and pressure build-up is provided by a relief valve, which should be installed on the water-outlet line as shown below.

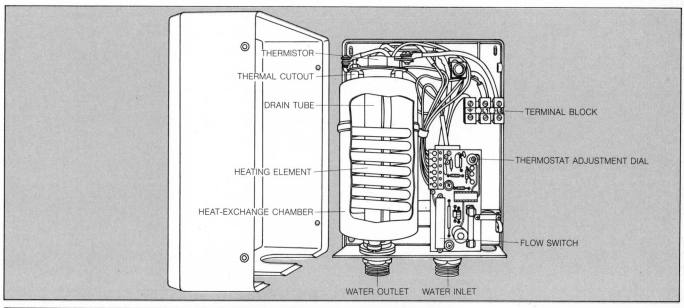

THERMISTOR
THERMAL CUTOUT
DRAIN TUBE
HEATING ELEMENT
HEAT-EXCHANGE CHAMBER
TERMINAL BLOCK
THERMOSTAT ADJUSTMENT DIAL
FLOW SWITCH
WATER OUTLET     WATER INLET

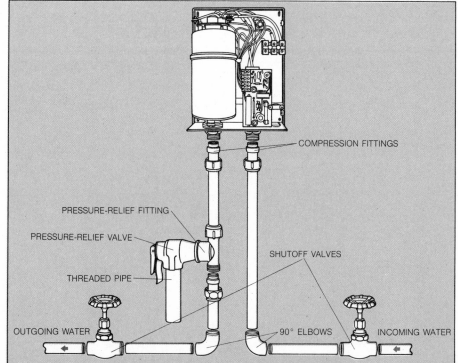

COMPRESSION FITTINGS
PRESSURE-RELIEF FITTING
PRESSURE-RELIEF VALVE
THREADED PIPE
SHUTOFF VALVES
OUTGOING WATER
90° ELBOWS
INCOMING WATER

**Plumbing connections.** A vertical length of copper tubing is first connected to the heater's water inlet with a two-piece compression fitting, consisting of a threaded nut and a soft brass ring. To make this connection, place the nut and then the ring onto the tubing, and push the tubing into the inlet as far as it will go; tighten the nut to the inlet, compressing the brass ring against the tubing. A shutoff valve is added near the heater, and the plumbing is then extended back to an existing water line with appropriate fittings and lengths of tubing.

A preassembled pressure-relief fitting, provided with the heater, is attached to the water outlet with a compression fitting, as above. Then a relief valve, rated to open at 150 pounds per square inch of pressure or 210° of temperature, is turned into the threaded horizontal opening in the pressure-relief fitting, and threaded pipe is added to the valve to channel released water downward. Finally, the outgoing water line is extended to the plumbing fixture or appliance. This line, too, needs a shutoff valve, so that the unit can be fully shut down for servicing. Before the final connections are made, however, the water flow must be adjusted as shown on page 112.

# Turning On Power and Water

**Wiring the heater.** Most electric heaters require a 240-volt power supply to carry the 30 to 40 amperes of current needed to heat the water, and must be wired directly to the main electrical panel. Run plastic-sheathed copper cable—No. 10 for 30 amperes, No. 8 for 40 amperes—from the main electrical panel to the heater, but leave the circuit unconnected at the main panel. At the heater, strip the cable sheathing and the ends of the two insulated wires; trim all three wires to the same length. Loosen the cable clamp at the lower right of the heater, and pull the cable through the opening.

Loosen the three screws on the lower part of the heater's electrical terminal block, and line up the ends of the wires with the corresponding terminals on the block—the bare ground wire with the ground terminal on the left, the insulated black and white wires with the terminals marked L-1 and L-2 (either wire can be connected to either terminal). Then forcefully push the ends of the three wires into the spring-loaded clamps in the terminals, and tighten the three screws to hold the wires securely in place. Do not make the connection at the main electrical panel until you have checked the plumbing for leaks and adjusted the water flow (*below*).

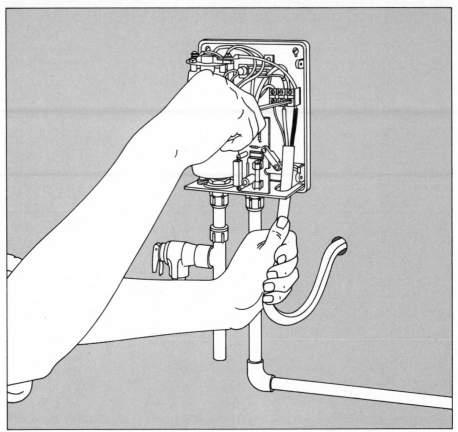

**Adjusting the water flow.** After the incoming water line is hooked up, place a 1-gallon container, such as a plastic milk bottle, under any convenient connection on the outgoing line, and set a pan underneath to catch any overflow. Open the shutoff valve on the incoming line fully, and have a helper turn on water to the new line at the nearest shutoff valve or at the main valve. Measure the time it takes for the container to fill; have your helper turn off the water.

Compare the measured rate of flow with the flow recommended for your particular installation. To adjust the flow, close the heater's incoming water valve a half turn and have your helper turn on the water; remeasure the rate of flow. Adjust it in this manner until the proper flow rate is reached. Then, leaving the incoming water valve in position, shut off the main water supply and connect the heater's outgoing water line to the plumbing fixture or appliance.

Turn on the main water supply again and check the new plumbing for leaks. If you are expert in electrical work, connect the 240-volt circuit to the main electrical panel; otherwise, have a professional electrician make this connection. Turn on power to the heater. Set the water temperature by turning the screw on the thermostat adjustment dial and checking the heated water with a thermometer. Attach the cover to the heater with the four screws provided.

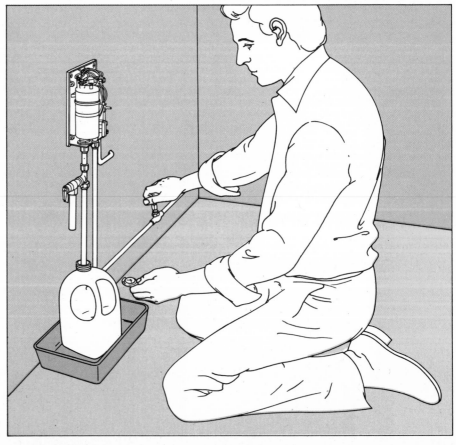

# A Timer That Saves Current

Conventional electric water heaters are undiscriminating consumers of electricity, keeping a tankful of water hot even when no one is home to use it. But such profligacy is readily avoidable: An inexpensive timer switch, similar to the kind that turns lights on and off, will program the water heater to warm up its contents when you need it and shut itself off for the rest of the day.

Water-heater timers are available at department or hardware stores. Be sure to select a timer that has voltage and watt capacity matching those of your water heater. Most electric heaters will require a 240-volt, 10,000-watt timer, the type shown on these pages. To install the timer, you will need only the most basic wiring skills and a few electrical tools and materials.

The most important tool is a voltage tester to check that the current is off before you touch any hot wires. You will also need plastic-sheathed indoor cable to reach from the top of the heater to the spot where you plan to mount the timer—on a stud or joist as close to the heater as possible. Select cable that has a ground wire and that matches the gauge of the cable connecting the heater to the main electrical panel of the house.

To cut and strip the cable and its wires, you will need a cable stripper and a wire stripper. Buy two cable clamps to match the cable gauge and a few cable staples to anchor loose cable to studs or joists once the installation is complete.

To operate the timer, you must set the clock to the actual time of day, and then fasten trippers at selected hours around its rim. As the clock rotates, the trippers flip a switch that shuts the current on and off at the times you have specified. You will get the best results if you set the timer to switch on one hour before hot water is needed in the morning, and off as soon as it is no longer being used; then on again one hour before evening use and off for the night. For maximum savings, try to hold the total "on" time to no more than three hours.

Insulation will make the system even more efficient; with an insulated water heater, the water in the tank will stay warm enough to heat up fairly quickly. If your water heater is not insulated already, it is best to put the insulation in place before you install the timer. You should also add foam insulation to all exposed pipe runs.

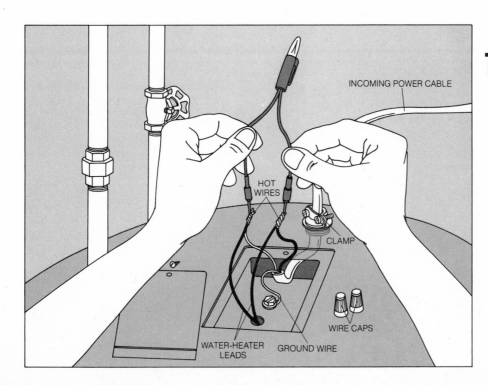

INCOMING POWER CABLE

HOT WIRES

CLAMP

WATER-HEATER LEADS

GROUND WIRE

WIRE CAPS

**1 Testing a circuit for safety.** Cut off the current to the water heater at the service panel, either by flipping the circuit breaker or by unscrewing the fuse. Then, at the water heater, unscrew and remove the small metal plate covering the lead wires that connect the heater to the wires of the incoming power cable. Twist off the wire caps that join the two sets of wires, taking care not to touch any exposed wire. Set the tester probes against the twisted ends of the hot wires. Then set one probe against each twisted wire in turn while holding the other probe against the ground wire. None of the three tests should produce a light in the voltage tester. If the tester lights up even once, you have disconnected the wrong circuit at the service panel. Continue to test until you find the correct circuit.

When the power to the water heater has been shut off, detach the ground wire and untwist the heater leads from the incoming power wires. Loosen the cable clamp around the power cable, and pull the cable free of the heater.

**2** **Preparing the timer for installation.** Open the door of the timer box, and press the spring lock at the top of the box with one hand to release the timer clock. Hold the lock down while you lift and tilt the rim of the clock with the other hand, freeing it from the box (*right, top*). Punch out the two knockout holes in the bottom of the box, and insert two cable clamps through them; be sure the heads of the clamping screws face front, for ease in tightening later. Anchor the clamps to the inside of the box with star nuts, tapping each point of the star with a nail set and a hammer to tighten the nut against the box (*right, bottom*).

Pry up the door latch so that it will be in the proper position to slip into the door slot. Then use the keyhole openings at the back of the box to fasten the box to a stud or joist near the water heater and within reach of the incoming power cable.

Replace the timer clock in the box by sliding it into the top of the opening and snapping it in place. Pull off the protective plastic cover that insulates the timer terminals.

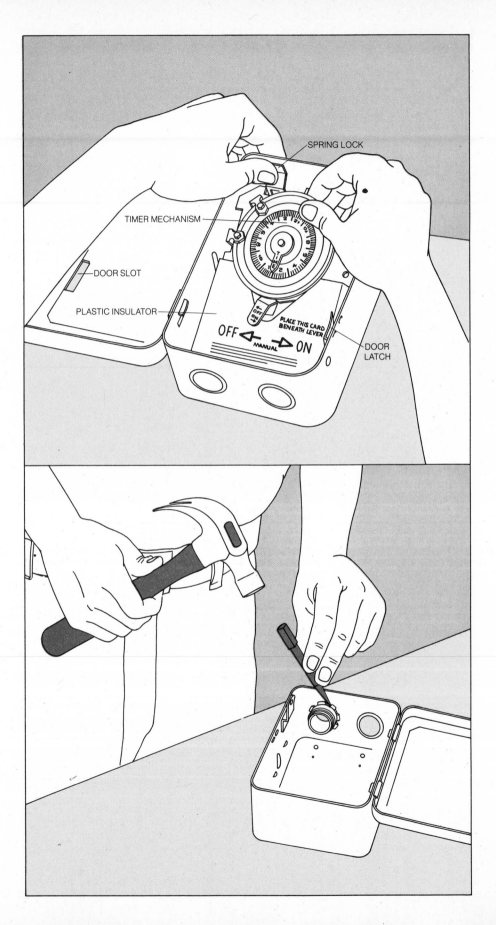

SPRING LOCK

TIMER MECHANISM

DOOR SLOT

PLASTIC INSULATOR

PLACE THIS CARD BENEATH LEVER

OFF ◄— MANUAL —► ON

DOOR LATCH

**3 Connecting the power cable.** Cut a piece of cable long enough to reach from the timer box to the leads on the top of the water heater. Strip approximately 4 inches of sheathing from each end of the cable and 1 inch of sheathing from the end of each of the wires inside the cable. Then push one end of this length of cable up into the timer box through the cable clamp beneath the terminals marked LOAD. Keep pushing until the cable sheathing bumps against the box, and tighten the screws of the cable clamp. Then slide the ends of the two hot wires up into the slots behind the terminals, and tighten the terminal screws.

Fasten the wires of the incoming power cable to the terminals marked LINE in the same manner, tightening the cable clamp around the cable as before. Then crimp the two ground wires together with a wire cap, adding a short length of connector wire, and fasten the connector wire to the ground screw (inset).

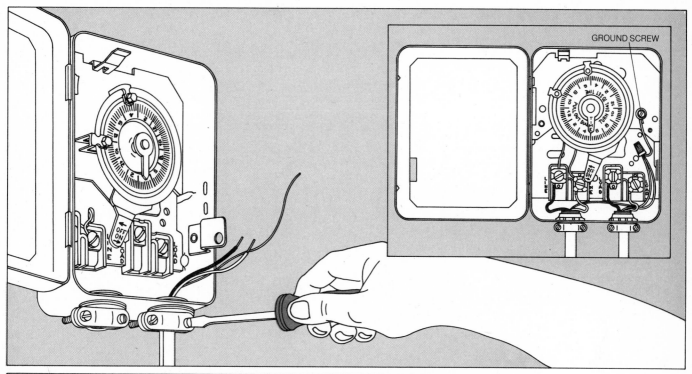

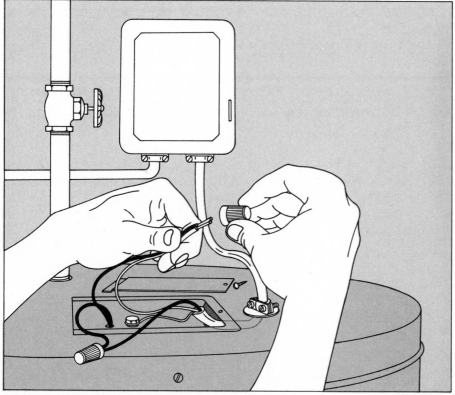

**4 Connecting the timer to the heater.** Run the remaining cable end into the access hole at the top of the water heater, and anchor it at the access hole with a cable clamp. Connect the two cable wires to the water-heater leads with wire caps; twist the caps clockwise until they are tight. If bare wire remains exposed, remove the cap, cut the wires and twist the cap back on. Connect the ground wire to the ground screw. Then replace the metal plate that covers the wires.

Set the timer clock and trippers according to the manufacturer's instructions. Push the timer switch to ON and replace the protective plastic cover that insulates the terminals. Close the timer door and put a padlock through the latch. Before turning the current back on, be sure all loose cable is anchored to solid surfaces with cable staples.

# Backyard Alternatives to Store-bought Electricity

Almost since the day that Thomas Edison invented the light bulb, the generation and distribution of electricity has mainly been the province of large, centralized utility companies. In recent years, however, concern for the environment and fast-rising prices of fossil fuels have prompted an intense search for alternative power sources that are clean, cheap and renewable. And the search is beginning to pay off.

Most promising among the new alternative power sources are sunlight, water power and wind. Each is widely available in virtually unlimited quantities, presents no environmental hazards in its use and in most cases can be employed even in remote areas not served by conventional utility lines. Although these natural sources are free, the equipment necessary to harness them is expensive. But recent advances have begun to bring down the cost and in some cases—particularly at the homes and homesteads of individuals who place a high economic value on self-sufficiency—photovoltaic cells *(page 123)*, water wheels *(pages 124-125)* and windmills are proving practical.

Although the use of wind for power harkens back reassuringly to the tower-mounted pinwheels that have stood for generations in the barnyards of the Great Plains, modern windmills bear scant resemblance to their rural forebears *(page 122)*. With their many blades, traditional windmills can spin efficiently with the lower-speed winds that blow close to the ground, so their towers rarely top 30 feet. Modern wind machines, which have fewer blades in most cases, are designed for the higher-speed winds that blow unobstructed above houses and trees. To harness these brisker airstreams, modern windmills are situated atop towers two or more times as high as those bracing the pinwheel designs.

Today's wind-catching propellers are sleekly aerodynamic and sometimes take surprising shapes, resembling gigantic interlocked croquet wickets or the bowed blades of an enormous egg-beater *(pages 119-120)*. In fact, even the name has changed: Manufacturers now prefer the fancy name wind energy conversion system, or wind plant, to windmill, a venerable term that recalls the first use of the machines as grinders of corn and grain. But for all of the changes, the guiding principle remains the same: Onrushing airstreams spin a propeller and the resulting rotation produces electrical current.

Within limits, the stronger the wind, the greater the quantity of free electricity produced. Moreover, the usable power in the wind builds up at a rate far greater than proportionate increases in the speed of the wind itself. Wind power experts call this principle the "cubed law" of wind velocity: Doubling the wind speed yields an eightfold ($2^3$)—or cubed—increase in potential electrical power.

The limits to this rule are reached only when winds grow strong enough to endanger the generating equipment itself—a possibility forestalled in modern machines by designs that slow or stop them when wind velocity nears 40 miles per hour. Some models automatically change the pitch of their blades, some deploy weights to slow the blades and some tilt the entire top of the tower to swing the blades out of the dangerous winds.

Although the wind is free, you will have to expend considerable effort researching your locale before making a yes-or-no decision on erecting a wind plant. First, you must check local laws to determine whether a tower 60 or more feet tall is allowed in your area. Then you must test for a full year to be sure that winds averaging at least 10 miles per hour blow reliably at your site. In order to monitor your winds, you must estimate the height of any nearby hills, trees and buildings; the propellers of a wind plant have to rise at least 60 feet above the ground and 30 feet above any obstacle within 300 feet. Next, rent or buy a wind odometer *(page 118)* that will telescope to the planned height of your wind plant.

During the year of testing, also tally your consumption of electricity—in kilowatts—by keeping monthly utility bills. If you are erecting the wind machine at a remote site that has no utility service, you will have to estimate the amount of current you will need *(pages 8-9)*.

If utility lines are not available at your site, you should have them run in unless the cost is wildly prohibitive. You then will be able to interconnect your wind plant, an arrangement with definite advantages. For one, a utility-connected system requires neither a bank of batteries to store power for windless days, nor a device called an inverter for changing battery-stored, direct current to alternating current for use in the house—items that add thousands of dollars to the cost of installation.

Another advantage comes into play when the winds blow hard, generating more kilowatts than you can readily use. By law, utility companies in the United States must buy any excess power that is fed into their transmission lines by independent producers.

Once you have studied the wind at your site and calculated the number of kilowatts you require, you are ready to investigate different generating systems. Begin by contacting the Rocky Flats Wind Systems Program in Golden, Colorado, the federal program that tests and rates wind machines. Then check dependable manufacturers for literature, technical data and prices.

In comparing wind machines, pay particular attention to what the equipment companies call a power curve, actually a graph that represents the monthly power output of each specific machine at different wind speeds. A less expensive wind plant may produce less power and, over its expected 20-year life, yield fewer kilowatts per dollar than a machine with a higher price tag.

When considering cost, remember to figure in deductions from federal and state taxes for energy equipment that is powered by renewable resources.

Once you make a decision, ask the manufacturer for the name of the best installer in your area. While you may want to test your winds and plan your installation on your own, erecting a 60-foot tower and attaching a heavy, cumbersome wind machine is a project best left to a professional. In fact, some manufacturers will not honor their warranties if the equipment is not installed by approved personnel.

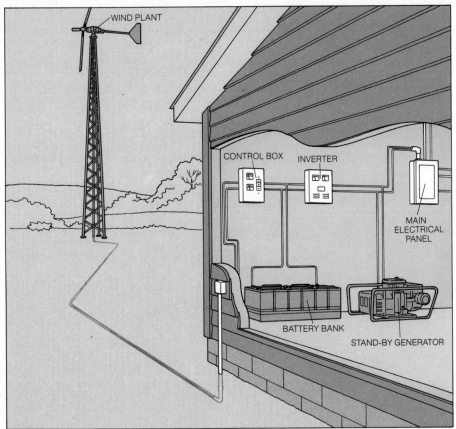

WIND PLANT

CONTROL BOX  INVERTER

MAIN
ELECTRICAL
PANEL

BATTERY BANK

STAND-BY GENERATOR

**An independent wind system.** Wind-driven blades atop a tower turn a generator to create direct current and send it through a buried cable to the house. There, a control box—containing electronic circuitry—regulates the voltage to match that of a bank of storage batteries. The control box also displays on its dials the voltage and amperage being produced by the system, and protects the batteries against excessive overcharging or discharging.

Beyond the batteries, an inverter converts the direct current to alternating current that can be fed into a conventional main electrical panel, which contains circuit breakers to protect the house wiring from overloads or short circuits. An optional, standby, gasoline- or diesel-powered generator can be added into the wiring system, as here, between the inverter and the service panel; should a long, windless period deplete the batteries severely, the generator can be run to power the house's lights and appliances.

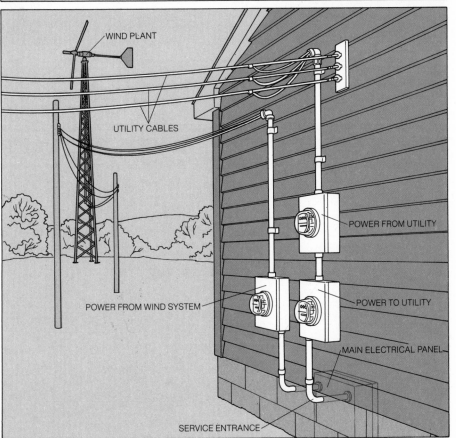

WIND PLANT

UTILITY CABLES

POWER FROM UTILITY

POWER FROM WIND SYSTEM

POWER TO UTILITY

MAIN ELECTRICAL PANEL

SERVICE ENTRANCE

**A utility-connected wind system.** When wind speed reaches 10 miles per hour, the tower-mounted anemometer on this utility-connected wind system closes a switch, sending current to the specially designed generator, which can operate as a motor to twirl the blades into motion. Then the wind takes over to drive the generator, which is engineered to run at a near-constant speed—no matter how fast the wind blows—in order to produce alternating current identical to that of the utility.

Cables connecting the wind machine to the service entrance of the house are interrupted by a meter that indicates the amount of electricity produced from the wind. Two other meters, each of which can turn in only one direction, are installed along the line that connects the utility company transformer to the main electrical panel; one records the amount of current returned to the utility by the wind system, the other records the current drawn from the utility to cover periods when the generator is producing less power than the residence requires.

# Assessing the Site for Wind-power Potential

**Finding heights of nearby obstacles.** Step off, to within a yard or two, the distance between an obstacle to wind flow and the proposed site of the wind plant. In the illustration at right, this measurement is labeled A. Standing at the site, stretch your arm horizontally and hold a ruler vertically in your hand. Have a helper measure the distance—B—between your eye and the ruler in inches. Without moving your arm, measure the apparent height—C—of the obstacle on the ruler. To find the height of the obstacle, convert all measurements to inches, then multiply the measurement for A by the one for C, and divide the result by the measurement for B.

**Determining average wind speed.** Buy or rent a wind odometer that includes a cup anemometer (which resembles a small pinwheel), a counter and a telescoping mounting pole. Mount the cup anemometer on the pole, raise it to the planned height of the wind plant and secure it against toppling; most poles come equipped with guy wires that attach to stakes.

Set the counter to zero on the first day of each month (*inset*). Record the total on the last day of each month. Each revolution shown on the counter represents the passage of 1/60 mile of wind, so divide each month's total by 60 to determine the number of miles of wind that have blown by the site. Divide the number of miles by the number of hours in a month—720 for a 30-day month—to determine the average monthly wind speed in miles per hour. At the end of the year, total the monthly averages and divide by 12 to find the average yearly wind speed at your site.

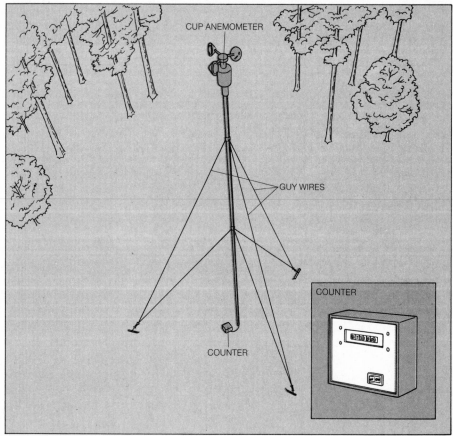

CUP ANEMOMETER

GUY WIRES

COUNTER

COUNTER

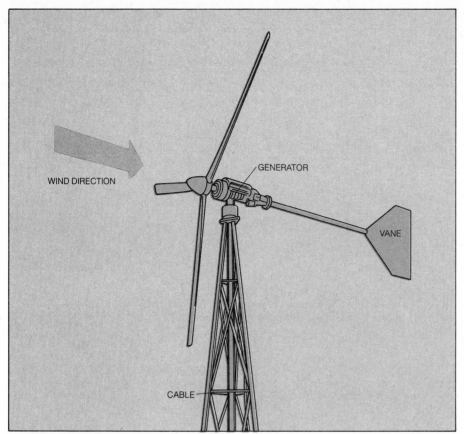

WIND DIRECTION

GENERATOR

VANE

CABLE

## Modern Power Producers

**An upwind horizontal-axis machine.** Called upwind because the tail or vane of the machine keeps the spinning blades oriented toward the wind, and horizontal because the axis of the propeller is parallel to the ground, this type of wind machine most nearly resembles the traditional windmill. The working parts are bolted to a steel tower; a cable connected to the generator runs down a tower leg. One virtue of this traditional design is that it has the longest history of reliability; a disadvantage is that it swings around with every change of wind direction, which wears down the machine's yaw bearings relatively quickly. For this reason it is best used where wind direction is steady.

**A downwind horizontal-axis machine.** Called downwind because breezes pass the generator before striking the blades, this machine forgoes a vane; the sleek shape of the generator housing and blades keep it correctly oriented. Although simpler in design than its upwind cousin, this model oscillates before finding the wind direction and is best used where wind blows mainly from one direction. The power cable is housed within the hollow mast that supports the machine.

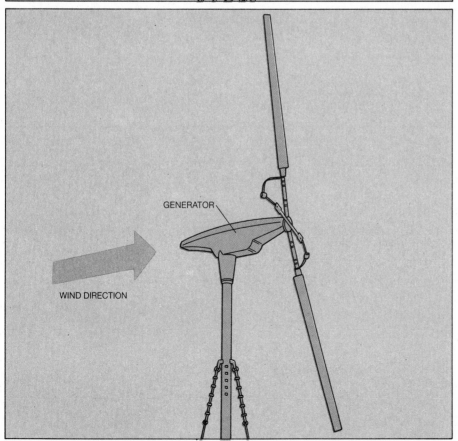

GENERATOR

WIND DIRECTION

**A vertical-axis pivoting-blade machine.** A vertical-axis machine, such as the one shown at right, has blades that spin around an axis perpendicular to the ground. It can exploit wind blowing from any direction without the entire machine having to pivot, an advantage in places where the wind direction changes frequently. Instead, special pivoting hardware between the struts and the blades in this model shifts the angle of the blades slightly with each rotation in order to present the broad surfaces of the blades to the breezes for as long as possible.

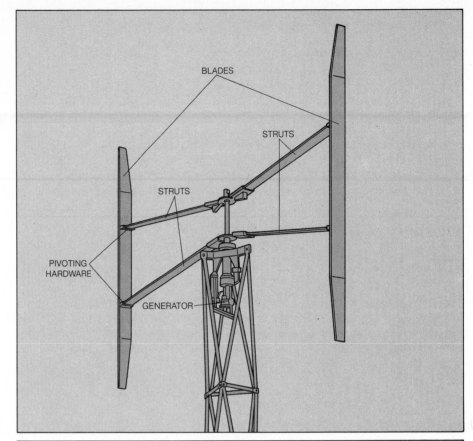

**A vertical-axis rotor machine.** This high-capacity vertical-axis model operates in much the same way as the one above except that its blades are fixed, unable to change pitch as they revolve around the center pole. Moreover, it cannot be started by the wind alone; it must be used in a utility-connected system so that electricity can spin the rotor until the wind takes over. It produces alternating current only. Unlike conventional wind machines that perch atop a tower, this one sits on a squat base, generally no more than 6 feet tall. The blades, sometimes as long as 70 feet, extend high into the air to catch the wind. Because the bottom ends of the blades are so close to the ground, this machine is best suited to flat areas or hilltops.

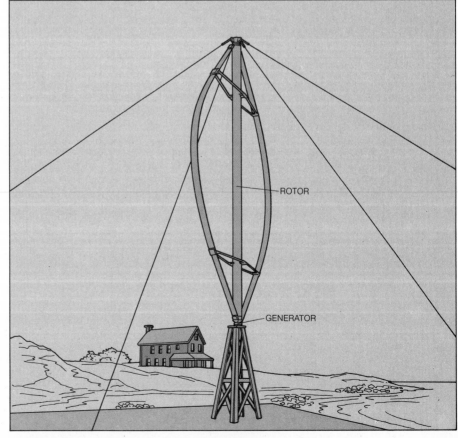

# How Professionals Install a Wind Machine

Some homeowners, chiefly those experienced in working at heights, have successfully erected their own wind-plant towers. Usually, however, this hazardous task is best left to practiced installers. A professional team has the expertise to accomplish the task safely, and it has ready access to costly equipment that can speed and simplify the job.

Before raising a wind-system tower, professionals assess soil conditions at the site and pour concrete footings to anchor the tower and brace it against anticipated stresses. The footings may have to be massive. The structure shown here, a 60-foot tower that is designed to stand without guy wires, needs three reinforced concrete piers, each 2 feet across and 8 feet deep.

Once the concrete of the footings has cured, the assemblers bolt steel struts together into tower sections 10 to 20 feet tall. The bottom section is hoisted atop the piers by crane and then held steady while workers bolt it to anchors in the piers *(below, left)*. Next, team members scale the bottom section, securing themselves to it with harnesses so that their hands will be free, and grasp the second section as it is delivered by crane *(below, center)*. After bolting the second tower section into place, the installers climb up to position and secure the third section. This climb, catch and clinch sequence is repeated until the team has topped the tower.

Lastly, the crane operator raises the generator, which itself may weigh several hundred pounds; workers on the ground guide steadying ropes attached to the payload so that a sudden gust does not send it crashing into the tower. At the tower peak *(below, right)*, a lone installer grasps the generator, guides it into place and bolts it down. Just before descending the tower, the installer connects an electrical cable to the generator, and then clamps the cable to a tower leg on the way down.

To join the wind plant to the house, the installers dig a trench, lay in electrical cable rated for underground use and cover it over. For an independent system, an experienced electrician makes connections at the control box, the batteries, the main panel in the house and, if needed, the inverter. For a utility-connected system, utility employees come out after the cable is laid and interconnect it with the existing wiring system in the house; at the same time they install additional meters or safety devices called for in their locality.

STEADYING ROPES

# Windmills Helped Win the West

During the 1800s, settlers moving west across the Great Plains found huge expanses of wind-swept prairie so arid that it could barely support coarse, tough-rooted buffalo grass, much less nurture crops, cows and families. Luckily, the settlers quickly realized that the same drying winds that were their bane could be their boon: Harnessed to drive pumps, the winds would dependably raise precious water from the depths—night and day, day in and day out.

The means of harnessing the western winds was the windmill, but perfecting a machine capable of withstanding the wild winds of Idaho and the tornadoes of Kansas was a challenge that required tenacity and ingenuity. At first, American manufacturers borrowed from the Dutch, trying to adapt the huge canvas sails of the familiar European windmill. But no matter how well sewn, the cloth sails could not stand the ferocity of western American weather. Crucial to the design of an Americanized windmill was a device that could both head the machine into the wind on a day of steady blowing and protect it from running wild in a gale. One early machine that fit the bill—a self-regulating windmill—was put together by an inventor named Daniel Halladay.

Halladay's breakthrough eschewed large sails in favor of dozens of thin wooden blades, similar to the slats of a venetian blind. Fixed to a ring around the hub, rather than to the hub itself, the slats stood upright before gentle breezes. In strong winds, however, the slats tilted away from the wind, so that they closed into a cylindrical shape that resembled an open-ended barrel; dangerous gusts could roar harmlessly around and through.

Simpler than the Halladay machines were the pinwheel fans turned out by Mast, Foos and Company, of Springfield, Ohio (below). Instead of using thin wooden slats, the Mast fans consisted of seven large fixed steel blades. A side vane, or rudder, perpendicular to the fan's axle, swung the entire mechanism around edgewise when a heavy wind blew. When the wind's pressure lessened, a hanging weight counteract-ed the effect of the side vane to pull the fan back into the wind's path.

During the second half of the 19th Century, manufacturers marketed a vast array of models and makes of windmills. At the World's Columbian Exposition, held in Chicago in 1893, companies exhibited dozens of models that not only pumped water, but shucked corn, ran lathes and powered sewing machines. Wrote one newspaper reporter, "Each manufacturer claimed some superiority. Here a wheel would open to get more wind or shut against too much; one mill would go swiftly in the lightest breeze, another would work slowly in a hurricane."

By the turn of the century, one of the largest sellers of windmills was Sears, Roebuck and Company. Its offerings included the high-capacity Suburban model, for those who wanted running water in their homes as well as their barnyards, and the Direct-stroke Steel machine, a durable, medium-capacity model, which could spin for years with little maintenance in remote pastures. While many manufacturers painted their own names on the rudder to advertize the product, Sears charmed its customers by offering to put the own-er's name on the rudder free of charge.

Between 1830 and 1935, an astonishing 6.5 million windmills were produced by American factories. In time, however, demand declined. Municipal water systems began replacing individual wells in towns, and in the 1930s the Rural Electrification Act brought economical, utility-generated power to most of the farms and ranches in the western states. Wind-driven water pumps gave way to more reliable electric ones. Now, although several of the original manufacturers are still in business selling kits to modern homesteaders, the windmill water pump no longer occupies a central place on the American farm. And long past are the days when pioneer farmers would drawl that the prairie is "no place for a woman unless she can keep a sod house tidy, shoot a snake and climb a windmill."

**A 19th Century windmill.** The seven curved blades of this iron-bladed machine not only provided the pumping power to water the cattle, but filled a storage tank on the top floor of the main house as well, ensuring a plenitude of running water for the then-rare indoor plumbing fixtures on this prosperous prairie farm.

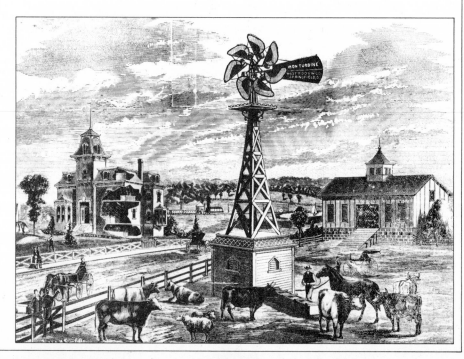

# Electricity from Sun and Silicon

With the help of thin, smooth, glasslike wafers, clean and inexhaustible sunshine can be put to work making electricity—with neither noise nor motion nor effluent. Individually, the wafers, known as photovoltaic cells, produce very small amounts of electricity; grouping cells into panels, and then joining the panels into larger assemblages called arrays, is necessary to yield enough electricity to run house lights and appliances.

Although there are some new experimental materials for cells, most now are made of silicon, the common element of sand. Despite the abundance of the raw material, however, the laborious manufacturing process renders the cells—and hence the solar-produced electricity—extremely expensive. Amortizing the cost of a solar array over its projected 20-year life span yields electricity for about five times the cost of utility-produced power and about twice that of wind-generated kilowatts. The cost decreased tenfold between 1970 and 1980 and is likely to continue to decline. But for now, this cleanest, quietest method of producing electricity at home remains as uneconomical as it is appealing.

**The workings of a solar cell.** The energy in sunlight, penetrating a solar cell, dislodges electrons from billions of silicon atoms. Because the cell—usually about 3 inches in diameter—is divided into two layers, each treated with a different chemical additive to induce opposite electrical charges in either side of the cell, the freed electrons rise to the top. There, the electrons are routed by scores of tiny metallic bus bars to larger bus bars reaching to the edge of the cell. A wire connected to the largest bus bar allows electrons to flow to a metal base plate on the bottom of the cell, where they can rejoin the electron-poor atoms. This moving file of electrons is electric current. As long as the sun shines and the plastic coating that protects the cells holds up—usually about 20 years—the cells will route electrons up, around the circuit, and back down in a never-ending stream.

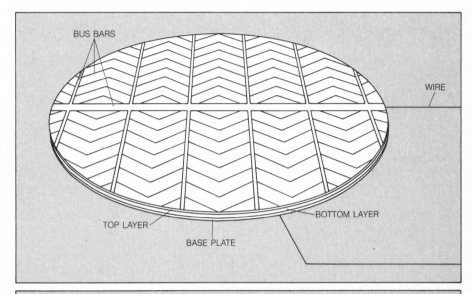

BUS BARS

WIRE

TOP LAYER

BOTTOM LAYER

BASE PLATE

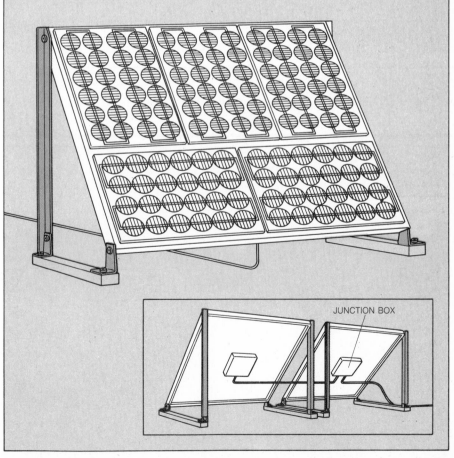

JUNCTION BOX

**A solar array.** Five solar panels, each composed of 24 disklike photovoltaic cells, are joined within a sturdy frame to form a solar array. The output of a solar array is equal to the yields of all of the individual cells added together; the array shown at right, for example, will produce about 54 volts and 1.3 amperes—or about 70 watts—enough to power a large black-and-white television set. If more power is required, several arrays can be wired together through junction boxes on the backs of the panels (*inset*).

To exploit all available sunlight, arrays are mounted on south-facing racks angled to within 15° of the local latitude. Like wind- and water-powered generating systems, photovoltaic systems also may be interconnected with utility lines or designed to stand alone, relying on a bank of batteries to store power for use during the night and on cloudy days.

# Making Electricity from a Mountain Stream

If you live in a mountainous area and a brook flows swiftly through your property, you can tap its energy with the miniature equivalent of a hydroelectric power station. A relatively small and simple system based on a water-driven turbine and a small electric generator can supplement current from your local utility. With a larger turbine and generator, you can free yourself from commercial power sources altogether. Large systems are costly, but where utility-supplied power is not available and the only alternative is a gasoline- or diesel-powered generator, a home system can prove competitive.

The ability of a stream to produce electric power depends on two factors: flow, the technical term for the volume of water the stream carries; and head, the drop in elevation along the portion of the watercourse where the system will be built. Where little head is available—as in a river flowing through flatlands—an expensive, complex dam is essential to back up the stream and create an artificial difference in water level. A mountain stream, on the other hand, often supplies sufficient head for a home hydroelectric system and needs only to be dammed with a few boulders or logs to channel sufficient flow into the system's inlet pipe, called the penstock.

Before planning a water-powered generating system, you must calculate both the head and the flow available from your stream. (Remember, however, that even streams running through private property are subject to laws regarding conservation and water rights; check with the local water resources agency before undertaking a hydroelectric project.) The easiest way to measure the flow of a stream no more than 7 feet wide is to interrupt it with a temporary plywood weir (opposite). The flow of a larger stream may be listed at a local U.S. Geological Survey office or, in Canada, at the municipal or district office responsible for natural resources. If not, you can hire a civil engineer to take flow readings. In any case, you will need monthly flow figures for an entire year to be sure that the water level does not drop too low for generating purposes during the dry season. Also check to see that the stream is not subject to destructive flooding.

To measure the head that your stream can provide, you will need a long measuring pole marked off in 6-inch increments, a second pole exactly 5 feet long, a 50-foot steel tape, a hand-held sighting level and a helper. In general, you will need at least 50 feet of head to make home electrical generating practical.

Once you have exact figures for the flow and the head, a simple formula allows you to calculate the amount of power available in the stream: Multiply the head measurement, in feet, by the flow, in cubic feet per minute; then divide the result by 708 to get the theoretical yield, in kilowatts of electric power. Because no system, no matter how efficient, will enable you to tap completely the theoretical potential of a stream, the calculated yield should be at least twice the number of kilowatts you need for your household (page 8).

You will need the head and flow figures—as well as your kilowatt requirements and a detailed description of the topography surrounding your stream—when you begin shopping for a turbine and generator. Turbine manufacturers—most of them small firms in the western United States—generally sell their equipment in complete systems, each tailored to the customer's site and needs. The system will include the turbine and either an alternating- or a direct-current generator. For systems that are to be interconnected with public utility lines, the supplier will also include hardware and safety devices to ensure that the electricity produced is compatible with that delivered by the utility (page 116).

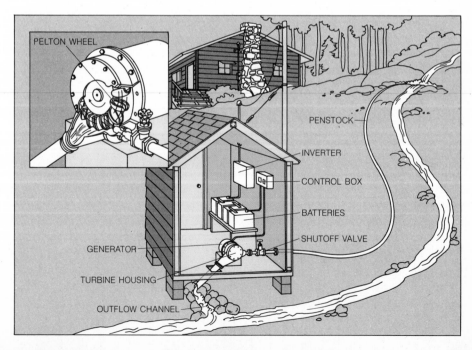

**A hillside power plant.** In this typical small-scale hydroelectric system, boulders placed in the stream bed make the water pool so that it covers the water intake, which is screened to keep out fish and debris. A penstock of inexpensive plastic pipe carries the water downhill into a prefabricated shed serving as the powerhouse, where it spins the turbine, here a many-armed type known as a Pelton wheel (inset). A large valve between the penstock and the turbine regulates the flow and allows the system to be shut down completely for maintenance. Beyond the turbine, water is carried out of the powerhouse into an outflow channel.

In this small system, independent of a local utility, the turbine drives a DC generator, which supplies current to a large bank of storage batteries. From the batteries, the direct current flows through an inverter, which changes it to alternating current, and then into the house. A voltage regulator, located in a control box, prevents the batteries from being overcharged.

## Measuring Head and Flow

**Using poles and a level.** Have a helper hold a pole, marked at 6-inch intervals, next to the stream 50 feet up the slope from the proposed powerhouse location. Sight the pole through a hand-held sighting level supported atop a 5-foot pole at the powerhouse site. Subtract the height sighted on the calibrated pole from the 5-foot height of the support to find the change in elevation. Move the level and its support to the location of the calibrated pole, and move the calibrated pole 50 feet upslope. Take a second reading and again subtract it from 5 feet. Repeat the process, working up the slope until the calibrated pole is directly opposite the inlet location. Add all of the measured changes in elevation (*inset*) to calculate the total head.

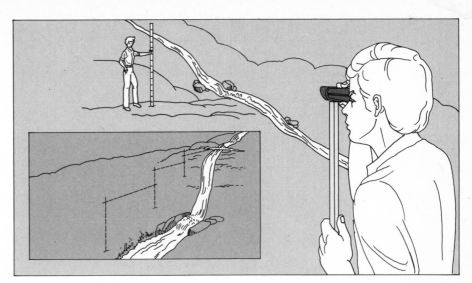

## Measuring Flow

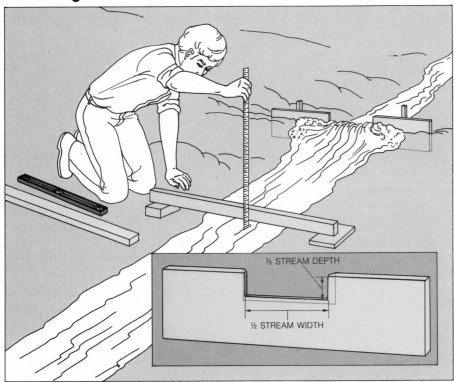

½ STREAM DEPTH

½ STREAM WIDTH

| Depth in inches | Flow value |
|---|---|
| 1 | 0.4 |
| 1.5 | 0.7 |
| 2 | 1.1 |
| 2.5 | 1.6 |
| 3 | 2.1 |
| 3.5 | 2.6 |
| 4 | 3.2 |
| 4.5 | 3.8 |
| 5 | 4.5 |
| 5.5 | 5.2 |
| 6 | 5.9 |
| 6.5 | 6.6 |
| 7 | 7.4 |
| 7.5 | 8.2 |
| 8 | 9.1 |
| 8.5 | 10.0 |
| 9 | 10.8 |
| 9.5 | 11.7 |
| 10 | 12.7 |
| 10.5 | 13.7 |
| 11 | 14.6 |
| 11.5 | 15.6 |
| 12 | 16.7 |
| 12.5 | 17.7 |
| 13 | 18.8 |
| 13.5 | 19.9 |
| 14 | 21.0 |
| 14.5 | 22.1 |
| 15 | 23.3 |
| 15.5 | 24.4 |
| 16 | 25.6 |
| 16.5 | 26.8 |
| 17 | 28.0 |

**1** **Taking depth readings with a weir.** Trim a 4-by-8-foot sheet of ¾-inch plywood to a length about 1 foot greater than the stream is wide and to a width about 8 inches greater than the stream is deep. Cut a rectangular notch in one of the long sides, about half the width and depth of the stream; bevel the notch with a rasp, leaving a ⅛-inch, squared-off ledge (*inset*).

Jam the weir into the stream bed, with the notch upward and the bevel facing downstream. Drive 2-by-2 stakes just downstream of the weir to brace it, and pack mud and gravel around the edges upstream to complete the seal. With the weir properly placed, the entire flow of the stream will pass through the notch. If the water overflows the notch, widen it or make it deeper. Lay a 2-by-4 board across the stream 5 feet upstream of the weir. Using a carpenter's level placed atop a 6-foot length of 2-by-4, adjust the board with pieces of scrap wood so that its top is even with the top of the weir. With a yardstick, measure the distance between the top of the board and the surface of the backed-up water. Subtract this distance from the depth of the notch in the weir to find the depth of the water roiling over the notch. Then use the table at right to calculate the water flow.

**2** **Calculating the flow.** Find the water depth over the weir notch (*Step 1*) in the first column of the chart above. Read across to the corresponding flow value in the second column. Multiply the flow value by the width of the notch, in inches; the result is the total flow of the stream, in cubic feet of water per minute.

## Picture Credits

*The sources for the illustrations in this book are shown below. The drawings were created by Jack Arthur, Roger Essley, William.J. Hennessy Jr., John Jones, Dick Lee, John Martinez and Joan McGurren.*

Cover: Fil Hunter. 6: Fil Hunter. 11-15: Wagner/Design. 18-25: Frederic F. Bigio from B-C Graphics. 26-31: Arezou Katoozian from A and W Graphics. 33-37: Walter Hilmers Jr. from HJ Commercial Art. 38: Fil Hunter. 41-47: John Massey. 49-57: Stephen Turner. 59-63: Adsai Hemintranont. 65-73: Eduino J. Pereira from Arts and Words. 74-77: John Massey. 79-87: Elsie J. Hennig. 89-93: Arezou Katoozian from A and W Graphics. 94, 95: Elsie J. Hennig. 96: Fil Hunter. 98-109: Frederic F. Bigio from B-C Graphics. 111-115: Stephen Turner. 117-121: Frederic F. Bigio from B-C Graphics. 122: from *Wind-Catchers, American Windmills of Yesterday and Tomorrow,* © 1976 by Volta Torrey, published by The Stephen Greene Press, courtesy Smithsonian Institution. 123: Frederic F. Bigio from B-C Graphics. 124, 125: Elsie J. Hennig.

## Acknowledgments

The index/glossary for this book was prepared by Louise Hedberg. The editors wish to thank the following: Greg Amonette, Pinsor Energy Corporation, Maston Mills, Mass.; Todd Bayer, Bethel, Vt.; Harold S. Boxer, Federal Solar Products Co., Edison, N.J.; James C. Brown and M. Robert Vincell III, Riteway Manufacturing Company, Inc., Weyers Cave, Va.; Paula Cartoun, The Tankless Heater Corporation, Greenwich, Conn.; Greg Clemmer, Solarex Corporation, Rockville, Md.; Ned Coffin, Enertech, Norwich, Vt.; CYRO Industries, Woodcliffe Lake, N.J.; John Darrow, Arlington, Va.; John Deahl, President, Solar Usage Now, Inc., Bascom, Ohio; John and Judy Dietrick, McLean, Va.; Russ Doran, Tumac Industries, Inc., Colorado Springs, Colo.; Michael E. Dorsey, Falls Church, Va.; Gould B. Flagg Jr., Texxor Corporation, Omaha, Neb.; Steve Hill, Preferred Sales, Millersville, Md.; Duke Jeffries, Associated Air Systems, Springfield, Va.; Robert Kiley, The Tankless Heater Corporation, Matamoras, Pa.; Dr. Jo Kohler, KLR Engineering, Keene, N.H.; Stan Lee, Carrier Automatic Equipment Sales, Alexandria, Va.; Ray Lewis, HI SQUARE Inc., West Palm Beach, Fla.; Tim Maloney, One Design, Winchester, Va.; Phil Metcalfe, UNR Rohn, Peoria, Ill.; Monegan, Ltd., Gaithersburg, Md.; Dan New, Canyon Industries, Deming, Wash.; Craig Nyman, Solar Works, Inc., Washington, D.C.; Potomac Electric Power Company, Residential Energy Services Department, Washington, D.C.; Donald Reece, Enertech, Norwich, Vt.; Layne Ridley, Passive Solar Industries Council, Alexandria, Va.; Robert P. Shapess, Climate Control Division, Singer, Auburn, N.Y.; William Shurcliff, Cambridge, Mass.; Small Farm Energy Project, Center for Rural Affairs, Hartington, Neb.; Dennis Tremblay, Kalwall Corporation, Solar Components Division, Manchester, N.H.; Andrew A. Turlington, PARCO, Inc., Clinton, Md.; Harold and Marcia Wallace, Energy Saver Shades, Springfield, Va.; Weather Energy Systems, Inc., Pocasset, Mass.; Michael Wilson, Vic's Supply, Falls Church, Va.; Henry Wolhandler, Contemporary Systems, Inc., Walpole, N.H.; Andrew Zaug, Hot Stuff Designer, La Jara, Colo. The editors also thank Robert Cox and Ann Miller, writers, for their help with this volume.

# Index/Glossary